# Artificial Intelligence for Cognitive Modeling

This book is written in a clear and thorough way to cover both the traditional and modern uses of artificial intelligence and soft computing. It gives an in-depth look at mathematical models, algorithms, and real-world problems that are hard to solve in MATLAB. The book is intended to provide a broad and in-depth understanding of fuzzy logic controllers, genetic algorithms, neural networks, and hybrid techniques such as ANFIS and the GA-ANN model.

**Features**

- A detailed description of basic intelligent techniques (fuzzy logic, genetic algorithm, and neural network using MATLAB)
- A detailed description of the hybrid intelligent technique called the adaptive fuzzy inference technique (ANFIS)
- Formulation of nonlinear models like analysis of ANOVA and the response surface methodology
- Variety of solved problems on ANOVA and RSM
- Case studies of above-mentioned intelligent techniques on the different process control systems

This book can be used as a handbook and a guide for students of all engineering disciplines, operational research areas, and computer applications, and for various professionals who work in the optimization area.

## Chapman & Hall/CRC Internet of Things: Data-Centric Intelligent Computing, Informatics, and Communication

The role of adaptation, machine learning, computational Intelligence, and data analytics in the field of IoT Systems is becoming increasingly essential and intertwined. The capability of an intelligent system is growing depending upon various self-decision-making algorithms in IoT Devices. IoT based smart systems generate a large amount of data that cannot be processed by traditional data processing algorithms and applications. Hence, this book series involves different computational methods incorporated within the system with the help of Analytics Reasoning, learning methods, Artificial intelligence, and Sense-making in Big Data, which is most concerned in IoT-enabled environment.

This series focuses to attract researchers and practitioners who are working in Information Technology and Computer Science in the field of intelligent computing paradigm, Big Data, machine learning, Sensor data, Internet of Things, and data sciences. The main aim of the series is to make available a range of books on all aspects of learning, analytics and advanced intelligent systems and related technologies. This series will cover the theory, research, development, and applications of learning, computational analytics, data processing, machine learning algorithms, as embedded in the fields of engineering, computer science, and Information Technology.

*Series Editors:*

**Souvik Pal**
Sister Nivedita University, (Techno India Group), Kolkata, India

**Dac-Nhuong Le**
Haiphong University, Vietnam

--------------------------------------------------------------------------

**Security of Internet of Things Nodes: Challenges, Attacks, and Countermeasures**
*Chinmay Chakraborty, Sree Ranjani Rajendran and Muhammad Habib Ur Rehman*

**Cancer Prediction for Industrial IoT 4.0: A Machine Learning Perspective**
*Meenu Gupta, Rachna Jain, Arun Solanki and Fadi Al-Turjman*

**Cloud IoT Systems for Smart Agricultural Engineering**
*Saravanan Krishnan, J Bruce Ralphin Rose, NR Rajalakshmi and N Narayanan Prasanth*

**Data Science for Effective Healthcare Systems**
*Hari Singh, Ravindara Bhatt, Prateek Thakral and Dinesh Chander Verma*

**Internet of Things and Data Mining for Modern Engineering and Healthcare Applications**
*Ankan Bhattacharya, Bappadittya Roy, Samarendra Nath Sur, Saurav Mallik and Subhasis Dasgupta*

**Energy Harvesting: Enabling IoT Transformations**
*Deepti Agarwal, Kimmi Verma and Shabana Urooj*

**SDN-Supported Edge-Cloud Interplay for Next Generation Internet of Things**
*Kshira Sagar Sahoo, Arun Solanki, Sambit Kumar Mishra, Bibhudatta Sahoo and Anand Nayyar*

**Internet of Things: Applications for Sustainable Development**
*Niranjan Lal, Shamimul Qamar, Sanyam Agarwal, Ambuj Kumar Agarwal and Sourabh Singh Verma*

**Artificial Intelligence for Cognitive Modeling: Theory and Practice**
*Pijush Dutta, Souvik Pal, Asok Kumar and Korhan Cengiz*

# Artificial Intelligence for Cognitive Modeling

## Theory and Practice

Pijush Dutta
Souvik Pal
Asok Kumar
Korhan Cengiz

CRC Press
Taylor & Francis Group
Boca Raton London New York

CRC Press is an imprint of the
Taylor & Francis Group, an **informa** business

A CHAPMAN & HALL BOOK

MATLAB® is a trademark of The MathWorks, Inc. and is used with permission. The MathWorks does not warrant the accuracy of the text or exercises in this book. This book's use or discussion of MATLAB® software or related products does not constitute endorsement or sponsorship by The MathWorks of a particular pedagogical approach or particular use of the MATLAB® software.

First edition published 2023
by CRC Press
6000 Broken Sound Parkway NW, Suite 300, Boca Raton, FL 33487-2742

and by CRC Press
4 Park Square, Milton Park, Abingdon, Oxon, OX14 4RN

*CRC Press is an imprint of Taylor & Francis Group, LLC*

© 2023 Pijush Dutta, Souvik Pal, Asok Kumar and Korhan Cengiz

*Library of Congress Cataloging-in-Publication Data*
Names: Dutta, Pijush, author. | Pal, Souvik, author. | Kumar, Asok
(Computer scientist), author. | Cengiz, Korhan, author.
Title: Artificial intelligence for cognitive modeling : theory and practice / Pijush Dutta, Souvik Pal, Asok Kumar, Korhan Cengiz.
Description: First edition. | Boca Raton : Chapman & Hall/CRC Press, 2023.
| Series: Chapman & Hall/CRC internet of things. Data-centric
intelligent computing, informatics, and communication | Includes bibliographical references and index. | Summary:
"This book is written in a clear and thorough way to cover both the traditional and modern
uses of Artificial Intelligence and soft computing. It gives an in-depth look at mathematical models, algorithms, and
real-world problems that are hard to solve in MATLAB. The book is intended to provide a broad and
in-depth understanding of fuzzy logic controllers, genetic algorithms, neural networks, and hybrid techniques such as
ANFIS and the GA-ANN model"-- Provided by publisher.
Identifiers: LCCN 2022050123 (print) | LCCN 2022050124 (ebook) | ISBN
9781032105703 (hbk) | ISBN 9781032461397 (pbk) | ISBN 9781003216001 (ebk)
Subjects: LCSH: Fuzzy expert systems. | Artificial intelligence--Industrial
applications--Case studies. | Neural networks (Computer science) | Soft computing.
Classification: LCC QA76.76.E95 D88 2023 (print) | LCC QA76.76.E95
(ebook) | DDC 006.3/3--dc23/eng/20221215
LC record available at https://lccn.loc.gov/2022050123
LC ebook record available at https://lccn.loc.gov/2022050124

ISBN: 978-1-032-10570-3 (hbk)
ISBN: 978-1-032-46139-7 (pbk)
ISBN: 978-1-003-21600-1 (ebk)

DOI: 10.1201/9781003216001

Typeset in Palatino
by SPi Technologies India Pvt Ltd (Straive)

# Contents

# *Preface*

This book presents both the traditional and modern aspects and applications of artificial intelligence and soft computing in a clear and highly comprehensive style. It provides an in-depth analysis of mathematical models, algorithms, and demonstrations of real-life complex problems in MATLAB. This book contains 15 chapters altogether. This book contains six case studies on industries such as liquid flow process control, machinery process industries, chemical industry, biomedical systems, manufacturing processes, and renewable energy. We have organized this book in such a manner so that the main objective is the realization of intelligent systems using methodologies. This book is unique for its contents, clarity, precision of presentation, and the overall completeness of its chapters. Understanding logic requires a basic understanding of data structures and C. All the simulation results are obtained from the MATLAB code and SIMULINK. So, this book is useful both for people who are interested in engineering and for research scholars because of what it says and how it says it. This book has a lot of computer simulations that show how different intelligent control techniques were used in real-world case studies. The book is organized into 15 chapters as follows.

**Chapter 1** discusses the overview of intelligent systems and soft computing. This demonstrates the scope of their application in real-world problems with complex systems. It highlights the historical development and how it was polarized in different domains. This chapter addresses different software tools that enable decision makers to draw on the knowledge and decision processes of experts in making decisions. Many public accounting firms have put in place some kind of intelligent system to help auditors and make their work more efficient. The most commonly used intelligent systems in public accounting are neural networks, genetic algorithms, and fuzzy logic. Research has led to neural networks that can learn and process datasets roughly, as well as the genetic algorithm for systematic random search and the fuzzy logic controller for roughly estimating reasoning. This chapter also talks briefly about how the book is set up in light of these tools and methods.

**Chapter 2** focuses on the practical application of fuzzy logic. Humans and machines differ in that human reasoning is ambiguous, imprecise, and fuzzy, whereas machines and the computers that power them are predicated on binary logic. Fuzzy logic is a technique for increasing the intelligence of machines by giving them the ability to think in a fuzzy style, similar to how humans do. A fuzzy controller is a knowledge-based controller in which fuzzy logic is used to represent knowledge and logical inference. In the present research, the implementation of the fuzzy set and its application in the field of fuzzy logic controllers is described. Beyond these types of membership functions, fuzzification and defuzzification techniques are also described.

**Chapter 3** presents a practical approach to neural network models. Biological systems are often compared to intelligent systems when examining how humans carry out control functions or make decisions. Many academics are getting more interested in deep learning because intelligent systems are becoming more popular and can adapt to new situations. Most of the time, neural networks are used to build systems with multiple stages that test learning algorithms by controlling their weights and biases. This chapter talks about the basics of neural networks. It explains the main parts of the network and how they work together. With the use of examples, it also covers several kinds of learning algorithms and activation functions. In common implementations, these principles are explained in depth.

**Chapter 4** introduces a novel type of classical algorithm for intelligent search and optimization problems that mimics the biological evolutionary process. After a brief introduction to the components of objective functions, objectives, and how to solve any objective function using optimization, different categories of optimization techniques with suitable examples are followed by different steps of genetic algorithms and their application in different domains, depending on the characteristics of the constraints. The last part of the chapter talks about differential evolution, which is one of the most well-known combinational optimizations.

**Chapter 5** explains in detail the theoretical background of the adaptive neuro-fuzzy inference system (ANFIS) model. Depending upon the learning algorithm, the ANFIS model classification and characteristics are also explained graphically. Finally, overall steps are highlighted to estimate any train and test dataset.

**Chapter 6** explores four different types of machine learning approaches, including supervised learning, unsupervised learning, reinforcement learning, and inductive logic programming. Decision tree learning and variation space-based learning are heavily emphasized within the supervised class of machine learning. A sample classification issue is used to briefly introduce the unsupervised class of learning. Q-learning and temporal difference learning are examples of the reinforcement learning discussed in this chapter. The basic idea and motivation of inductive logic programming are illustrated in the last section.

**Chapter 7** analyzes three different bio inspired optimization techniques. Particle swarm optimization (PSO), flower pollination algorithm (FPA), and cuckoo search optimization (CSO) are used efficiently for estimating the drain and transfer characteristics of two model parameters of JFET (Model No: J112A N-channel JFET). In addition, we can successfully forecast the JFET's I-V characteristics, which reinforces the approach that has already been established. To perform the comparative study among these three algorithms, three criteria were chosen, namely RMSE, computational time, and convergence speed. According to the results, FPA outperformed the other two algorithms in terms of computational time, convergence speed, and RMSE by about 3.41% and 2.19% for drain and transfer characteristics, respectively.

**Chapter 8** discusses a neural network approach to determine the impact of epidemiological parameters that affect risk factors. Five elements in all, which are categorized as risk factors, have been taken into consideration. This neural network model supports understanding and evaluating the impact of these factors on the spread of COVID-19. Also, the model establishes a basis for understanding the effect of risk factors and vice versa. In this chapter, a total of 162 datasets are used, which contain the input parameters virulence, immunity, temperature, and populations; the output is risk value. The model response was cross-verified with the actual risk value and found to be satisfactory. In the second phase of work, the optimized conditions of epidemiological factors were found to give the fittest model of risk factor of the test dataset.

**Chapter 9** discusses the manufacturing of microscopic parts using the electrochemical discharge micro-machining technique ($\mu$-ECDM). During micro-channel cutting on silica glass ($SiO_2$+$NaSiO_3$), parametric effects on the material removal rate (MRR), machining depth (MD), and overcut (OC) have been proposed utilizing the applied voltage (V), pulse-on time (s), stand-off distance (SD), and mixed electrolyte concentration (wt%). The effectiveness of the model has been evaluated using a fuzzy logic based AI tool, and membership based on the magnitude of the input parameters resulting in the highest cutting depth with the most material removal at the shortest overcut has been determined using optimization. Results analysis further reveals that employing a mixed electrolyte on a cylindrical tool and computer-assisted subsystem movement for the $x$, $y$, and $z$ axes enhanced machining depth and surface quality.

**Chapter 10** focuses on two different ANFIS models: grid partition and subclustering algorithm, which are used for fast and best prediction of average localization error (ALE). Transmission range (TR), node density (ND), and anchor ratio (AR) and iterations (IT) are considered as features in a dataset for prediction of ALE. In experimental result it is seen that both the ANFIS models produce the same RMSE error of 0.069, but the correlation coefficient of grid partition outperforms the later ANFIS model.

**Chapter 11** discusses three optimization techniques: GA, PSO, and a hybrid algorithm of GA and PSO (HGAPSO). For a single diode model and a double diode model of a solar cell, the statistical result is correlated with the particle swarm optimization (PSO) and genetic algorithm (GA). A comparative study reveals that the upgraded rendition of the gray wolf optimization tool provides a more accurate model for predicting the ideal solar cell parameters with the fewest possible iterations. Hence we recommended HGAPSO as the best optimization tool for providing the perfect performance.

**Chapter 12** designs an evolutionary method for optimization problems of nonlinear systems. Because it is used in the biological field, a cylindrical tank is used here as a nonlinear example. Its inversion is caused by the way the method proceeds, making it feasible to expel the products without any squandering. The typical PID controllers' closed loop capabilities are assessed in order to govern the tank's level. Flower pollination algorithm (FPA) and bacterial foraging optimization (BFO), two naturally based optimization techniques, are deployed to enhance performance, as given in two segments overall: in the first part, the model is represented into two transfer functions, the ordinary first order function and first order with time delay; and in the second part, proper tuning of PID controller parameters of both the transfer functions is determine by BFO and FPA algorithms. Simulation results show that the flower pollination algorithm is better than BFO in terms of transient behavior, and that the proposed nature-based optimization techniques are better than the proportional integral controller.

**Chapter 13** talks about the process variables of the liquid flow system, which are often changed while the equipment is still running. Therefore, it appears that choosing the right level of organizational factors, or interaction effects, is key to gaining the best flow rate. The determination of the ideal linear combination quantities in the liquid flow rate mechanism is the main emphasis of the current study. Input parameters include liquid conductivity, pipe diameter, flow sensor output, and viscosity, while response parameters include flow rate determined by testing. The process was initially analyzed quantitatively using ANOVA. In the next step, a number of newly suggested metaheuristics, such as PSO, CSO, and hybrid CSPSO, are used to optimize the liquid process involved parametrically in order to maximize the responsiveness in parameters. The suggested approach was corroborated by the simulated outcomes, which also validated the test.

**Chapter 14** discusses a synthetic minority over-sampling technique (SMOTE) based deep neural network for early prediction of diabetes. We used 768 datasets with 17 attributes to do this research. There are 268 false, true, and positive samples in these datasets, and 500 positive samples. In the first stage, SMOTE is used to get rid of randomness, imbalance, and make the oversampled dataset more accurate and stable. In the second stage, balanced datasets are applied in a deep neural network regression model to train and test the datasets. In the last step, we used three different ways to measure how well our model worked: confusion matrix parameters, statistical parameters, and computational time.

**Chapter 15** explains how machine learning algorithms are used to collect evaluations from the Internet and classify them into six categories for the prediction of human behavior: walking, going upstairs, walking downstairs, sitting, standing, and laying. This chapter identifies the participants' behavior patterns and attempts to gain more insights. Seven different machine learning classifiers are employed in this work to determine the accuracy, exactness, recollection, and F1-score to identify the best model. Throughout the analysis, it is seen that the linear support vector machine (LSVM) shows average accuracy of about 97%, far better than the other methods.

We are sincerely thankful to the Almighty for supporting and standing with us at all times, whether it's good or tough times, and giving us ways to concede. Starting from the writing of the chapters till the finalization of the chapters, all the editors gave their contributions amicably, which is a positive sign of significant teamwork. The editors are sincerely thankful to all the members of CRC Press, especially Aastha Sharma and Isha Singh, for providing constructive input and allowing an opportunity to write this important book. We are equally thankful to the reviewers who hail from different places in and around the globe who shared their support and stood firm toward quality chapter preparation.

**Pijush Dutta**
**Souvik Pal**
**Asok Kumar**
**Korhan Cengiz**

# *Acknowledgment*

First of all, we would like to thank all our colleagues and friends for sharing our happiness at the start of this project and following up with their encouragement when it seemed too difficult to complete. We are thankful to all the members of CRC Press, especially Gagandeep Singh, Aastha Sharma, and Isha Singh, for giving us the opportunity to write this book.

Prof. Dutta is extremely grateful to his parents for their love, prayers, care, and sacrifices for educating him and preparing for his book project. He is thankful to his friends for their love, understanding, prayers, and continuing support to complete this book project.

Dr. Pal is grateful to his father, Prof. Bharat Kumar Pal, and mother, Smt. Tandra Pal, for their affection and constant support. He is also grateful to his grandmother Late Gita Rani Pal and grandfather Late Ajit Kumar Pal for their blessings. He is thankful to his beloved wife, Smita, and son, Binayak, for their love and encouragement. This book has been a long-cherished dream, which would not have been turned into reality without the support and love of these amazing people, who encouraged him with proper time and attention.

Dr. Kumar wishes to thank his parents and family members for their continuous support. He is also grateful to his best friends for their blessings and unconditional love, patience, and encouragement.

Dr. Cengiz would like to acknowledge and thank the most important people in his life, his parents and his partner for their support.

And above all, God is almighty.

**Pijush Dutta**
**Souvik Pal**
**Asok Kumar**
**Korhan Cengiz**

# Authors

**Pijush Dutta** received his B.Tech. and M.Tech. in electronics and communication engineering and mechatronics engineering from WBUT, India in 2007 and 2012, respectively, and is pursuing his PhD degree (submitted) from the Department of Electronics & Communication Engineering, Mewar University, India. Currently, he is working as an assistant professor and head of the Department of Electronics and Communication Engineering, Greater Kolkata College of Engineering & Management, Baruipur, India. He was the former head of ECE Department, Global Institute of Management Technology, Krishnagar, India from 2018 to 2022. His current research interests include sensors and transducers, nonlinear process control system, mechatronics systems and control, optimization, intelligent systems, Internet of Things (IOT), machine learning, and deep learning. He has 50+ publications in peer-reviewed journals and in national and international conferences. He also published 14 national and international patents so far. Prof. Dutta has authored three books in his credit. He is also an editorial and review board member of many peer-reviewed international journals.

**Souvik Pal** is an associate professor in the Department of Computer Science and Engineering at Sister Nivedita University (Techno India Group), Kolkata, India. Dr. Pal received his M.Tech and PhD degrees in the field of computer science and engineering from KIIT University, Bhubaneswar, India. He has more than a decade of academic experience. He is author or co-editor of more than 18 books from reputed publishers, including Elsevier, Springer, CRC Press, and Wiley, and he holds three patents. He has more than 75 publications in his credit in Scopus/SCI/SCIE journals and conferences. He is serving as a series editor for "Advances in Learning Analytics for Intelligent Cloud–IoT Systems," published by Scrivener-Wiley Publishing (Scopus-indexed); "Internet of Things: Data-Centric Intelligent Computing, Informatics, and Communication," published by CRC Press, Taylor & Francis Group, USA; and "Conference Proceedings Series on Intelligent Systems, Data Engineering, and Optimization," published CRC Press, Taylor & Francis Group, USA. He is the organizing chair of RICE 2019, Vietnam; RICE 2020, Vietnam; and ICICCT 2019, Tunisia. He has been invited as a keynote speaker at ICICCT 2019, Turkey, and ICTIDS 2019, 2021 Malaysia. He has also served as proceedings editor of ICICCT 2019, 2020; ICMMCS 2020, 2021; and

ICWSNUCA 2021, India. His professional activities include roles as associate editor, guest editor, and editorial board member for more than 100+ international journals and conferences of high repute and impact. His research area includes cloud computing, big data, Internet of Things, wireless sensor networks, and data analytics. He is a member of many professional organizations, including MIEEE; MCSI; MCSTA/ACM, USA; MIAENG, Hong Kong; MIRED, USA; MACEEE, New Delhi; MIACSIT, Singapore; and MAASCIT, USA.

**Asok Kumar** received his B.Tech and M.Tech degrees in radio physics and electronics from Calcutta University in 1997 and 1999, respectively, and received his PhD degree from Jadavpur University, 2007. Currently Dr. Kumar working as a dean of the Student Welfare Department at Vidyasagar University, West Bengal, India. He has more than 21 years of teaching and more than 15 years of research experience. Apart from teaching and research activity, he has more than 10 years of administrative experience as a HOD, dean of academics, NBA coordinator, principal of several reputed engineering colleges. Dr. Kumar's research includes data security, wireless communication, information theory and coding, computer networking, intelligent control, soft computing, optimization, and sensors and transducers. He has more than 90 research articles in his credit, published in national, international, and conference proceedings. He is a reviewer of more than 10 national or international journals. He has authored two books and seven book chapters in the field of data security and networking. He has been invited as a keynote speaker at many national level institutes and organizations. He is a member of various professional bodies like IE (India), IEEE, ISTE, IACSIT, and so forth.

**Korhan Cengiz** received the BSc degree in electronics and communication engineering from Kocaeli University and in business administration from Anadolu University, Turkey in 2008 and 2009, respectively. He took his MSc degree in electronics and communication engineering from Namik Kemal University, Turkey in 2011, and the PhD degree in electronics engineering from Kadir Has University, Turkey in 2016. Since September 2022, he has been associate professor in the department of Computer Engineering, Istinye University, Istanbul, Turkey. Since April 2022, he has been the chair of the research committee of University of Fujairah, United Arab Emirates. Since August 2021, he has been an assistant professor at the College of Information Technology in the University of Fujairah, UAE. Dr. Cengiz is the author of over 40 SCI/SCI-E articles published in journals including *IEEE Internet of Things Journal, IEEE Access, Expert Systems with Applications, Knowledge Based Systems*, and *ACM Transactions on Sensor Networks*, as well as five international patents, more than 10 book chapters, and one book in Turkish. He is editor of more than 10 books. His research interests include wireless sensor networks, wireless communications, statistical signal processing, indoor positioning systems, Internet of Things, power electronics, and 5G. He is the associate editor of *IEEE Potentials Magazine*; handling editor of *Microprocessors and Microsystems*, Elsevier; and associate editor of *IET Electronics Letters*, IET Networks. He also holds a guest editorial position in *IEEE Internet of Things Magazine*. He serves several reviewer positions for *IEEE Internet of Things Journal*, *IEEE Sensors Journal*, and *IEEE Access*. He serves in several book editorial positions for IEEE, Springer, Elsevier, Wiley, and CRC. He has presented 40+ keynote talks in reputed IEEE and Springer conferences about WSNs, IoT, and 5G. He is a senior member of IEEE and a professional member of ACM. Dr. Cengiz's awards and honors include the Tubitak Priority Areas PhD Scholarship, the Kadir Has University PhD Student Scholarship, Best Presentation Award in ICAT 2016 Conference, and Best Paper Award in ICAT 2018 Conference.

# Part A

# Artificial Intelligence and Cognitive Computing

*Theory and Concept*

# 1

## Introduction to Artificial Intelligence

## 1.1 Introduction

With the implementation of framework complexity and modernization of process plants, a new approach has been taken which is dependent on evaluation, generation, planning, and conveyance time. Modern process plants aim for expanded efficiency, better item quality, and developing benefits to stay focused on the global economy. Automation is urgent for prudent plant activity through productive strategies, proper energy utilization, improved safety, and waste minimization. Automation-based process industries must meet criteria that include increased product quality, reduced factor lead times, increased productivity, improved security, and finally decreased undesirable signs.

Control framework configuration is incredibly dominant by the measure of nonlinearity present inside the process. If a linear model is somehow influenced by the tiny effects of nonlinearity in the system, then to provide satisfactory output over the wide range of the linear process, we use some classical controllers. However, the presence of a significant amount of nonlinearity or disturbance makes the linear models ineffectual even away from the working point. To compensate for such nonlinearity and disturbance, a technique known as adaptive control strategies is used. To compensate the corresponding variations in the properties of the process, the adaptive control framework uses feed forward, feedback, or a combination of the two. An adaptive control technique requires adjustable controller parameters and a mechanism like gain scheduling of a linear controller in a minor nonlinearity-affected process plant. But for several nonlinearities affecting the working condition and fluctuating parameter variations in a process plant, we need to design a dynamic model to overcome such conditions.

The exceptionally nonlinear behavior and time-differing parameters of liquid flow process systems make these a benchmark for display and control of nonlinear processes. Nonlinear processes can be demonstrated utilizing three different ways in particular: mathematical demonstration, dependence on the first principles approach, and framework separation dependent on experimental input–output information [1, 2]. Dynamic modeling, also called "white box" modeling, utilizes laws of conservation of mass, physical realization, and chemical laws. Its performance is always dependent on the first guideline. White box models [3, 4] are physical models depending on thermodynamics and additional scientific conditions and design techniques for vital display, analysis, and control. White box model strategies are fuzzy logic controllers where dynamic models explained with optimistic assumptions—for example, immaculate blending and nonattendance of estimation clamor—do not delineate the genuine and sensible conduct of the process.

DOI: 10.1201/9781003216001-2

Nonlinear adaptive control frameworks (white boxes) are incapable of effectively measuring the process parameters of modern process plants.

To design a black box model doesn't require any knowledge of framework design. The model can be implemented with the help of prior information of input and output, which is why it is also called a statistical based model. To prepare the black box model requires an enormous number of input–output datasets. Since 1990, artificial neural networks (ANNs), genetic algorithms (GAs), support vector machines, reinforcement adaptation, deep machine learning, and so on have been the most well-known black box modeling techniques applied in a wide scope of nonlinear framework applications. The ANN-based model is inspired by organic neural systems, and it contains a lot of interconnected nonlinear handling components known as counterfeit neurons. The GA-based model is motivated by bio-enlivened operators such as mutation, crossover, and selection. ANN and GA have an incredible capacity to learn nonlinear elements of mind-boggling processes due to their intrinsically parallel and conveyed design. That is the reason ANN- and GA-based model procedures have been broadly uncovered for the process control industry.

Framework-recognizable dynamic "gray box" models can be structured utilizing exploratory information obtained from the process. The "gray box" method needs some understanding of the framework separated from the test information. To overcome the drawbacks of white and black box models, hybrid models have been presented [5]. Gray box (hybrid) models are a blend of material science–based and statistical-based models. Gray box modeling is incredibly exact for process control framework and performance improvement.

## 1.2 Intelligent Control

For different modern industrial activities, and even household appliances, the connections between imprecision, vigor, and organization of these strategies have turned into the inexorably rise of "intelligent systems" [6–8].

Since the 1970s, scientists have proposed many control systems which incorporate different stages, such as modeling, analysis, simulation, implementation, and verification [9, 10]. Most of these control strategies have found their way into practice but have not received significant attention. In 1977, Fu and Saridis [11] first introduced the concept of intelligent control. In early 1962, Zadeh [12] articulated the intelligent control theorem.

There is no formal or single definition of an intelligent control framework. An intelligent control framework ought to fulfil the famous Turing test, which can be briefly expressed as follows: a similar task is done by a human and a machine (or a program); if at any instant one can't recognize the machine and the human by looking at just their ideas, then this machine is said to be intelligent, and otherwise not. Nonetheless, intelligent control frameworks can be comprehensively portrayed as the utilization of human-made reasoning strategies to the structure and execution of automated control frameworks. Dr. Fu first presented the term "intelligent control" and started pondering them in the 1950s. At the underlying phase of the advancement of intelligent control, there were tight ties between human brainpower and programmed control. Advances in software engineering and operation research also added to intelligent control during the years that followed.

## 1.3 Expert Systems

An expert system (ES) is computer software created to solve complicated issues and offer decision-making capabilities similar to those of a human expert [13, 14]. This is accomplished by the system retrieving information from its knowledge base in accordance with user queries, utilizing reasoning and inference procedures.

The first ES, which was the first effective use of artificial intelligence (AI), was established in 1970 and is a subset of AI. By drawing on the knowledge that is kept in its knowledge base, it can solve even the most complicated problems like an expert. Like a human expert, the system aids in decision-making for difficult issues by using both facts and heuristics.

A conventional problem-solving system has both programmers and data structures encoded, whereas an expert system just has data structures hard-coded and does not have any problem-specific information encoded in the program. A user interface and an inference engine are included in the knowledge engineer software used to build the majority of expert systems [15, 16]. Building a knowledge base is the main obstacle in the creation of expert systems, and this gap is encouraged. Not all expert systems contain learning components that allow them to adjust to new circumstances or satisfy new demands. But each expert system shares the feature that, after being fully constructed, it will be evaluated and proved using the same real-world problem-solving scenario. These systems are created for a specific industry, like science, medicine, and so on.

## 1.4 Soft Computing Techniques

In a wide point of view, intelligent control frameworks underlie what is designated "soft computing." Intelligent systems are an integration of methodologies which provide the foundation of conceptual design. They integrate the trade-off of precision and certainty of traditional hard computing systems and the computation, reasoning, and decision-making of soft computing systems, as shown in Figure 1.1. Moreover, the major part of the basic intelligent system is reciprocal instead of focused [17, 18]. Progressively, these methodologies are additionally applied as a combination, alluded to as "hybrid."

To improve AI, soft computing archetypes and their combinations are used, but in computing processes AI also incorporates human expert knowledge. Their applications include but are not limited to controlling complex systems, predicting the unknown parameters of geological changes or the world economy, controlling industrial processes, and so on. Figure 1.2 represents the basic structure of an intelligent control system.

The main advantages of soft computing methodologies compared to the analytical methods are that they can learn from the framework, they are capable of mapping output from input information, and their search space is global instead of local.

The improvement of soft computing techniques has encouraged significant research enthusiasm since 2010, which is definitely not a solitary philosophy; rather, it is a combination of a few approaches, namely neural systems, fuzzy logic controllers, and genetic algorithms. Unlike conventional (hard) computing, soft computing always exploits instinct. The inspiration for applying the human instinct is that an enormous number of genuine issues cannot be understood by conventional computing techniques because of the way

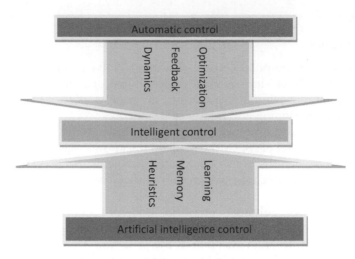

**FIGURE 1.1**
Basic diagram of a soft computing system.

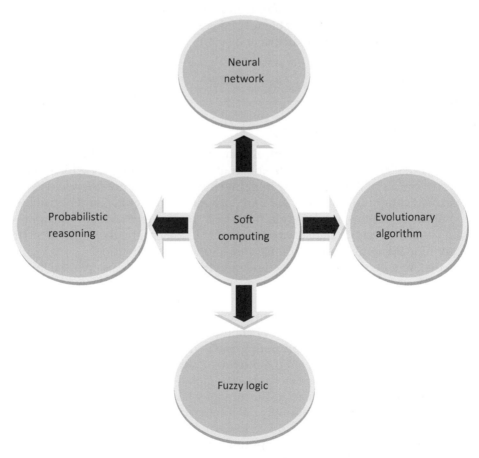

**FIGURE 1.2**
Basic structure of an intelligent control system.

**TABLE 1.1**

List of Basic Intelligent Systems with Their Advantages

| Name of the Basic Intelligent System | Advantage |
| --- | --- |
| Neural networks | Learning and approximation |
| Genetic algorithms | Systematic random search |
| Fuzzy logic | Approximate reasoning |

that possibly they are excessively mind-boggling to grasp or cannot be portrayed or classified by explanatory and definite models. The fuzzy set hypothesis depicts deficient ideas that are hard to figure out numerically. Meticulousness of decision and participation capacities for a given problem is the major issue for a fuzzy logic controller. Neural networks, a face of soft computing, frequently experience the ill effects of a moderate learning rate. This disadvantage renders neural systems not exactly appropriate for time-basic applications. The genetic algorithm is altogether raised on probabilistic as opposed to deterministic hunts. Table 1.1 shows the list of basic intelligent systems and their advantages.

Nonetheless, in soft computing, systems are accessorial as opposed to focused. More precisely, it is worthwhile to utilize the fuzzy logic control, neural systems, and genetic algorithm in a mix rather than straightforwardly. A fuzzy neural system is a hybrid intelligent tool incorporating frameworks of both fuzzy inference and neural systems, with the goal that their individual significances survive. Fuzzy-neural systems have a similar topology to feed-forward neural systems, through which they capture the inexact deduction attributes and loose data preparation capacity of fuzzy logic systems; then, they additionally have the quality of adjustment and speculation by taking in the calculations from neural systems. The genetic–neural strategy removes the extreme weakness of applying an unadulterated back-propagation to train the neural systems.

Soft computing methodologies mimic consciousness and cognition in important ways that differ from analytical approaches. They can learn from experience, universalize into domains where direct experience is lacking, and perform input-to-output mapping more quickly than inherently serial analytical representations offered by parallel computer architectures that simulate biological processes. The anticipated reduction in computational burden and subsequent rise in calculation rates that enable more robust control are the driving forces for such an extension. There is a large amount of literature on soft computing, both theoretical and practical. Section 1.4.1 introduces the concept of fuzzy logic as well as its applicability to various industrial processes. In Section 1.4.2, the justification as well as the rationale for the utilization of neural networks in various industrial applications is presented. The evolutionary computation is presented in Section 1.4.3. Section 1.4.4 is devoted to the integration of soft-computing methodologies commonly called hybrid systems. Finally, real-time systems are presented in Section 1.4.5.

### 1.4.1 Fuzzy Systems

Since about 1990, there has been much disagreement and spirited discussion around fuzzy logic. Zadeh, who is regarded as the father of the area, wrote the first article in fuzzy set theory, which is today regarded as the foundational study on the topic. In that work, Zadeh was subtly extending the idea of human approximate reasoning, which enables people to make wise choices based on the imperfect language data at their disposal [19–21]. Mamdani carried out the first implementation of Zadeh's concept in 1975, proving fuzzy

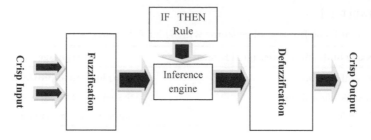

**FIGURE 1.3**
Functional diagram of fuzzy system.

logic control's (FLC's) applicability to a small model steam engine. There are, however, knowledge-based systems and information that cannot be described by conventional mathematical representations, noted in Section 1.1. Such pertinent subjective information is frequently disregarded by designers in the initial stages but is frequently used to assess designs in the final stages. Systems based on information and knowledge can be built using fuzzy logic. The so-called knowledge-based technique is considerably more in line with how people actually think and speak than is the conventional classical reasoning.

Fuzzy sets and the fuzzy logic hypothesis to change them into an important standard logic are shown in Figure 1.3. The target information exists in a numerical structure that is applied in some engineering points, and the transient learning exists in a phonetic structure which is for the most part impossible to systematize [22]. Fuzzy logic systems have been utilized for modeling and have been replicated in many real-time system problems as shown in Figure 1.4.

#### 1.4.1.1 Architecture of Fuzzy Logic Systems

Input variables: Crisp values changed into the fuzzification section.

Fuzzifier: The fuzzifier is transfiguring crisp surveyed facts into suitable etymological qualities.

Fuzzy inferencing: This is the brain of a FLC, which accomplishes the ideal control strategy after utilizing the skill of human knowledge through performing approximate consciousness.

Fuzzy rule base: This sends experimental information to a process domain expert.

Defuzzifier: This makes a non-fuzzy choice from a contingent control activity by the inference system.

Output variables: Defuzzification section changes output to crisp values. There are a number of books related to fuzzy logic [5, 19, 23–25].

### 1.4.2 Neural Networks

The neural systems attempt to replicate the natural utilities of the human mind. A neural system is a data processing model [26, 27] motivated by organic sensory systems. It is fundamentally comprised of an immense number of incredibly interconnected handling segments that are seeking to solve the specific problem. A neural network is exemplified for an exact application with the assistance of a learning technique which incorporates

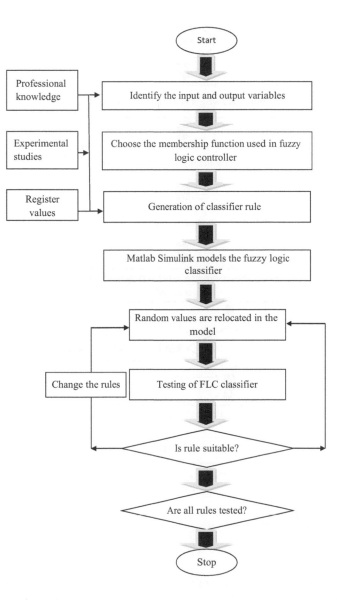

**FIGURE 1.4**
Basic framework of fuzzy logic controller.

changes to the synaptic associations between the neurons. The basic architecture and framework for back-propagation of an ANN model are shown in Figure 1.5.

### 1.4.2.1 Basic Architecture of Neural Network

Feed forward networks: In this model, data only proceeds in one direction [28]. Here feedback or loops are absent so the output is not affected by any intermediate layer. Feed forward neural networks in general are conventional forward systems that go to contributions with the output. They are exhaustively utilized in example acknowledgment.

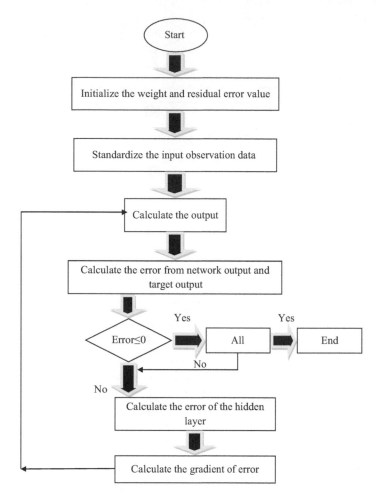

**FIGURE 1.5**
Framework for back-propagation of an ANN.

Feedback networks: Feedback systems enable signs to travel in two directions with the assistance of loops [29]. This method is extremely powerful and becomes exceedingly mind-boggling. Feedback network systems are self-persuaded, and their state is dynamic until it arrives at a steadiness point. They are utilized in the process control industry, forecast of ecological parameters, and so forth.

Although single layer feed forward networks, multilayer feed forward networks, and recurrent networks are the three major categories for neural network architectures, numerous different neural network structures have developed over time. Back-propagation networks, perceptrons, adaptive linear elements (ADALINEs), associative memory, Boltzmann machines, adaptive resonance theory, self-organizing feature maps, and Hopfield networks are a few of the well-known neural network systems.

To name a few, pattern recognition, image processing, data compression, forecasting, and optimization challenges have all been effectively solved using neural networks.

### 1.4.3 Genetic Algorithms

Genetic algorithms are stochastic inquiry enhancers that depend on the considerations of development and characteristic choice. GAs [30, 31] are characteristically parallel calculations, which repeat to exploit the present parallel supercomputers to speed up the streamlining task by a factor near the quantity of parallel specialists. GAs perform very well in blending multidimensional function areas, matters with discrete arrangement spaces, and nondifferentiable objective function. Global optimizers are generally autonomous of the solution domain, which is why they do not trouble the referenced imperatives of the optimization problems. Genetic algorithm are the most suitable to deal with such issues.

Because of the previously mentioned favorable circumstances of GAs, scientists have recently begun to utilize them for parameter enhancement of the procedure control industry. The basic framework of a genetic algorithm is shown in Figure 1.6.

Although binary coding of the problem parameters is used in the majority of GA simulations, real coding of the parameters has also been proposed and used. A wide range of scientific and engineering fields, including function optimization, machine learning, scheduling, and others, have found extensive applications for GAs, which have been theoretically and empirically demonstrated to enable robust search in a complicated space [22, 32].

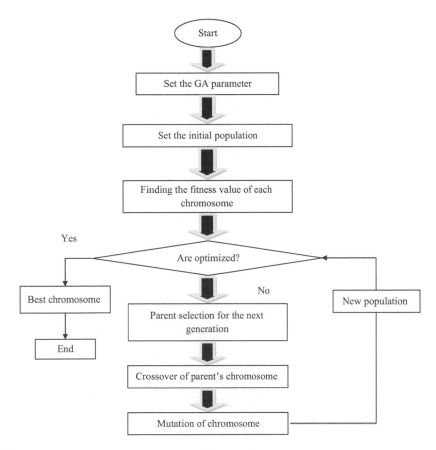

**FIGURE 1.6**
Basic framework of genetic algorithm.

### 1.4.4 Adaptive Neuro-Fuzzy Inference System

An adaptive neuro-fuzzy inference system (ANFIS) [33, 34] is an intelligent system (framework) which counterfeits a neural system dependent on the Takagi–Sugeno fuzzy inference framework, and was first exhibited in the 1990s [35]. In the versatile neuro-fuzzy framework, neural systems have viable learning calculations to improve tuning of the enrollment capacities and appropriate standards of fuzzy frameworks.

There are some basic aspects of these AI techniques which require better understanding, more specifically:

- No standards methods exist for transforming human knowledge or experience into the rule base and database of a fuzzy inference system.
- There is a need for effective methods for tuning the membership function (MFs) so as to minimize the output error measure or maximize performance index.

In this perspective, this novel architecture called ANFIS serves as a basis for constructing a set of fuzzy if–then rules with an appropriate membership function to generate the stipulated input–output pairs.

Fuzzy and neural systems are associated in their favorable circumstances and to recover their individual shortcomings. Neural networks present the computational attributes of the fuzzy frameworks and acquire the lucidity and understanding of framework portrayals. Numerous significant parameters of ANFIS support the framework (system) to perform serious tasks through fuzzy principles: precise adaptation, simplicity of actualization, fantastic clarification offices, and excellent speculation calibers. Most of the main applications are in system control. The applications appear in fields like data characterization, data analysis, decision-making, and defect recognition. The basic framework of ANFIS is shown in Figure 1.7.

ANFIS has been used in numerous time series research areas, including the application of ANFIS based on singular spectrum analysis for forecasting chaotic time series, chaotic time series prediction using improved ANFIS, fuzzy time series forecasting, developing a new method for predicting the trends of oil prices, predicting stock returns, and predicting financial volatility. ANFIS is ultimately superior to the other approaches.

### 1.4.5 Real-Time Systems

Real-time systems are frameworks operable to address issues as they arise in constant inserted programming frameworks, and issues that separate them from other programming frameworks. In a real-time system, the logical outcomes of the computations and the physical instant at which these results are produced both affect how the system behaves. Real-Time systems are categorized from a variety of angles, including aspects inside and outside the computer system. Real-time systems, both hard and soft, are given particular attention in this book.

In soft real-time systems, a missed deadline can result in a sizable loss, whereas in hard real-time systems, it is disastrous. Therefore, in these systems, predictability of system behavior is of utmost importance. The key highlights of the real-time frameworks are the convenient reaction prerequisite. Various programming structure procedures have been allotted for the distinguishing proof of parallel exercises and execution through learning of the planned framework. However, the vast majority of the methodologies depend on the

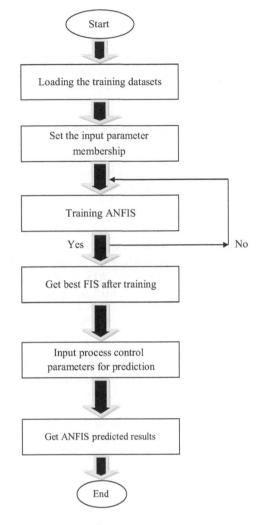

**FIGURE 1.7**
Flowchart of ANFIS models.

heuristics, in that exhaustive knowledge of the proposed framework, decision of programming language, and execution qualities of the extreme engineering assume key roles in the deconstruction of the framework into the simultaneous framework. Chemical concentration of a reaction chamber and all types of process control systems are typical examples of real-time systems.

## References

1. Wang F, Yang J-F, Wang M-Y, Jia C-Y, Shi X-X, Hao G-F, Yang G-F. Graph attention convolutional neural network model for chemical poisoning of honey bees' prediction. *Science Bulletin* 2020;65: 1184–1191.

2. Zhuang S, Hadfield-Menell D. Consequences of misaligned AI. *Neural Information Processing Systems* 2020;33: 15763–15773.
3. Loyola-Gonzalez O. Black-box vs. white-box: Understanding their advantages and weaknesses from a practical point of view. *IEEE Access* 2019;7: 154096–154113.
4. Zhang X, Rasmussen C, Saelens D, Roels S. Time-dependent solar aperture estimation of a building: Comparing grey-box and white-box approaches. *Renewable and Sustainable Energy Reviews* 2022;161: 112337.
5. Imamguluyev R. Application of fuzzy logic model for correct lighting in computer aided interior design areas. *Proceedings of the International Conference on Intelligent and Fuzzy Systems*, Springer;2020, pp. 1644–1651.
6. Nayak J, Naik B, Dinesh P, Vakula K, Rao BK, Ding W, Pelusi D. Intelligent system for COVID-19 prognosis: A state-of-the-art survey. *Artificial Intelligence* 2021;51: 2908–2938.
7. Polap D, Winnicka A, Serwata K, Kç sik K, Woźniak M. An intelligent system for monitoring skin diseases. *Sensors* 2018;18: 2552.
8. Punjabi SK, Chaure S, Ravale U, Reddy D. Smart intelligent system for women and child security. *Proceedings of the 2018 IEEE 9th Annual Information Technology, Electronics and Mobile Communication Conference (IEMCON)*. IEEE;2018, pp. 451–454.
9. Abdellatif T, Brousmiche K-L. Formal verification of smart contracts based on users and blockchain behaviors models. *Proceedings of the 2018 9th IFIP International Conference on New Technologies, Mobility and Security (NTMS)* IEEE;2018, pp. 1–5.
10. Jorissen F, Reynders G, Baetens R, Picard D, Saelens D, Helsen L. Implementation and verification of the IDEAS building energy simulation library. *The Journal of Building Performance Simulation* 2018;11: 669–688.
11. Passino, Kevin M. "Intelligent control: an overview of techniques." Perspectives in Control Engineering: Technologies, Applications, and New Directions, 2001, pp. 104–133.
12. Wilson C, Marchetti F, Di Carlo M, Riccardi A, Minisci E. Intelligent control: A taxonomy. *Proceedings of the 2019 8th International Conference on Systems and Control (ICSC)*, IEEE;2019, pp. 333–339.
13. Leo Kumar SP. Knowledge-based expert system in manufacturing planning: State-of-the-art review. *International Journal of Production Research* 2019;57: 4766–4790.
14. Medsker LR, Bailey DL. Models and guidelines for integrating expert systems and neural networks. *Hybrid architectures for intelligent systems*, CRC Press;2020, pp. 153–171.
15. Akanbi AK, Masinde M. Towards the development of a rule-based drought early warning expert systems using indigenous knowledge. *Proceedings of the 2018 International Conference on Advances in Big Data, Computing and Data Communication Systems (IcABCD)*, IEEE;2018, pp. 1–8.
16. Yelagandula SK. Designing an AI expert system 2020, pp. 1–23.
17. Falcone R, Lima C, Martinelli E. Soft computing techniques in structural and earthquake engineering: A literature review. *Engineering Structures* 2020;207: 110269.
18. Yan Y, Wang L, Wang T, Wang X, Hu Y, Duan Q. Application of soft computing techniques to multiphase flow measurement: A review. *Flow Measurement and Instrumentation* 2018;60: 30–43.
19. Arji G, Ahmadi H, Nilashi M, Rashid TA, Ahmed OH, Aljojo N, Zainol A. Fuzzy logic approach for infectious disease diagnosis: A methodical evaluation, literature and classification. *Biocybernetics and Biomedical Engineering* 2019;39: 937–955.
20. Hannan MA, Ghani ZA, Hoque MM, Ker PJ, Hussain A, Mohamed A. Fuzzy logic inverter controller in photovoltaic applications: Issues and recommendations. *IEEE Access*2019;7: 24934–24955.
21. Sharma S, Obaid AJ. Mathematical modelling, analysis and design of fuzzy logic controller for the control of ventilation systems using MATLAB fuzzy logic toolbox. *Journal of Interdisciplinary Mathematics* 2020;23: 843–849.
22. Ding X, Zheng M, Zheng X. The application of genetic algorithm in land use optimization research: A review. *Land* 2021;10: 526.
23. Alekseev VV, Vasilyev S. Application of fuzzy logic elements under the moisture supply evaluation in the plant-soil-air system. *CEUR Workshop Proceedings*, 2018, pp. 04–09.

24. Gupta M, Alareeni B, Akimova L, Gupta SK, Derhaliuk MO. Application of fuzzy logic data analysis method for business development. *Proceedings of the International Conference on Business and Technology*, Springer;2020, pp. 75–93.
25. Noor F, Tanjim MR, Rahim MJ, Suvon MNI, Porna FK, Ahmed S, Kaioum MAA, Rahman RM. Application of fuzzy logic on CT-scan images of COVID-19 patients. *Intelligent Information Systems and Intelligent Database Systems* 2021;14: 333–348.
26. Abiodun OI, Jantan A, Omolara AE, Dada KV, Mohamed NA, Arshad H. State-of-the-art in artificial neural network applications: A survey. *Heliyon* 2018;4: e00938.
27. Abiodun OI, Jantan A, Omolara AE, Dada KV, Umar AM, Linus OU, Arshad H, Kazaure AA, Gana U, Kiru MU. Comprehensive review of artificial neural network applications to pattern recognition. *IEEE Access* 2019;7: 158820–158846.
28. Ozanich E, Gerstoft P, Niu H. A feedforward neural network for direction-of-arrival estimation. *Journal of the Acoustical Society of America* 2020;147: 2035–2048.
29. Sam DB, Babu RV. Top-down feedback for crowd counting convolutional neural network. *Proceedings of the 32nd AAAI Conference on Artificial Intelligence*, 2018.
30. Mirjalili S. Genetic algorithm. *Evolutionary algorithms and neural networks*. Springer;2019, pp. 43–55.
31. Mirjalili S, Song Dong J, Sadiq AS, Faris H. Genetic algorithm: Theory, literature review, and application in image reconstruction. *Nature-Inspired Optimization Algorithms*;2020, pp. 69–85.
32. Mayer MJ, Szilágyi A, Gróf G. Environmental and economic multi-objective optimization of a household level hybrid renewable energy system by genetic algorithm. *Applied Energy* 2020;269: 115058.
33. Golafshani EM, Behnood A, Arashpour M. Predicting the compressive strength of normal and high-performance concretes using ANN and ANFIS hybridized with Grey Wolf optimizer. *Construction and Building Materials* 2020;232: 117266.
34. Moayedi H, Raftari M, Sharifi A, Jusoh WAW, Rashid ASA. Optimization of ANFIS with GA and PSO estimating α ratio in driven piles. *Engineering with Computers* 2020;36: 227–238.
35. Karaboga D, Kaya E. Adaptive network based fuzzy inference system (ANFIS) training approaches: A comprehensive survey. *Artificial Intelligence Review* 2019;52: 2263–2293.

# 2

## *Practical Approach of Fuzzy Logic Controller*

### 2.1 Introduction

Due to a degree of ambiguity in the factors that characterize the problem or the circumstances in which it arises, problems in the actual world frequently end up being complicated [1–3]. Despite being a tried-and-true method for dealing with unpredictability, a purposive approach can only be used in circumstances where the features are based on stochastic processes or situations where the recurrence of occurrences is solely decided by chance [4]. The truth is that there are issues, a sizable class of which have a nonrandom process as their defining characteristic of ambiguity. In this case, the uncertainty may be brought on by incomplete knowledge of the issue, information that is not entirely trustworthy, linguistic imprecision resulting from the problem's definition, or contradicting information from several sources [5, 6]. In these circumstances, fuzzy set theory shows tremendous promise for effectively addressing the problem's ambiguity. Vagueness is a synonym of fuzzy. A good mathematical method for addressing the uncertainty brought on by ambiguity is fuzzy set theory [7]. Fuzziness frequently appears while identifying handwritten letters or comprehending spoken language [8–12]. The many characteristics of the classical set and operation are described in this chapter.

### 2.2 Classical Set Properties and Operation

#### 2.2.1 Classical Set

1. Crisp bounds, meaning there is no question in the prescription or placement of the set's limits, are what constitute a classical set [13, 14].
2. The universe of discourse is the expanse of all knowledge that is obtainable for a specific issue. $X$ is typically used to represent the universe of discourse, while $x$ is used to represent each particle in the world.
3. The total number of components in the cosmos $X$ is called its cardinal number, which is denoted by $n_x$.
4. Collections of all the components inside a set are referred to as subsets, and sets are groups of the pieces within a set. The term "whole set" also refers to the collection of all components in the cosmos.
5. Null set ($\varphi$) is the set having no elements in the entire set ($X$). The power set $P(X)$ is a special set that includes all conceivable sets of $X$. The cordially of the power set, denoted by $n_p(x)$, is equal to $2^{n(x)}$.

DOI: 10.1201/9781003216001-3

Example: Consider a universe of discourse $X = \{a, b, c\}$. Calculate the cardinal number of $X$, power set, and cardinality of the power set.

Solution: cardinal number = number of elements in the universe = $n_x = 3$
Power set, $P(x) = \{\{\varphi\}, \{a\}, \{b\}, \{c\}, \{a, b\}, \{a, c\}, \{b, c\}, \{a, b, c\}\}$
Cardinality of power set = $n_p(x) = 2^{n(x)} = 2^3 = 8$

## 2.3  Properties of Crisp Sets

The qualities play a significant part in all mathematical processes since they may be used to determine how to resolve them [15–17]. The following are crucial characteristics of traditional sets:

1. Commutivity

$$A \cap B = B \cap A$$
$$A \cup B = B \cup A$$

2. Associativity:

$$A \cup (B \cup C) = (A \cup B) \cup C$$
$$A \cap (B \cap C) = (A \cap B) \cap C$$

3. Distributivity:

$$A \cup (B \cap C) = (A \cup B) \cap (A \cup C)$$
$$A \cap (B \cup C) = (A \cap B) \cup (A \cap C)$$

4. Idempotency:

$$A \cup A = A$$
$$A \cap A = \varphi$$

5. Identity:

$$A \cup \varphi = A \ \& \ A \cap X = A$$
$$A \cap \varphi = A \ \& \ A \cup X = X$$

6. Transitivity:

If $A \le B \le C$ then $A \le C$

In this case the symbol "$\le$" means contained in or equivalent to and "$<$" means contained in.

7. Involution:

$$\overline{\overline{A}} = A$$

De Morgan's law and the excluded middle laws are the other two significant exceptional qualities.

8. Law of excluded middle:

It represents the union of a set A and its complement
$$A \cup \overline{A} = X$$

9. Law of contradiction:

It represents the intersection of a set A and its complement.
$$A \cap \overline{A} = \varphi$$

## 2.4  Concept of Fuzziness

To explain fuzziness we take an example to make easier to understand the basic difference between Boolean logic and fuzzy logic [18].

### 2.4.1 Fuzzy Set

1. In a classical set, a constituent in the cosmos can abruptly and clearly go from being a member of a particular set to not being a member of that set. However, with a fuzzy set, this shift is slow since the fuzzy set borders are ill-defined and unclear [19].

2. A set called a fuzzy set has components with variable degrees of membership. A fuzzy set is a contained element that includes members of a set to variable degrees.

3. A set called a fuzzy set has components with variable degrees of membership. A fuzzy set is a contained element that includes members of a set to variable degrees. Fuzzy sets are usually denoted by $\tilde{A}$. $\tilde{A}$ maps elements of fuzzy set $\tilde{A}$ to a real numbered value on the interval 0 to 1. If an element in the universe, say $X$, is a member of fuzzy set $\tilde{A}$, then this mapping is given by $\mu_A(x)\ \varepsilon\ (0,1)$. The mapping is shown in Figures 2.1 and 2.2 for a typical fuzzy set [20].

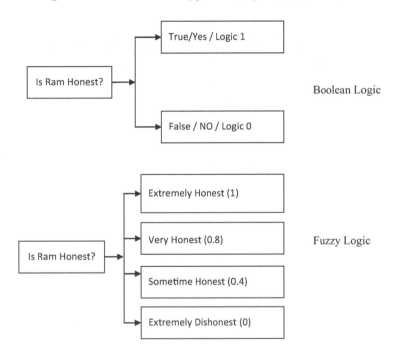

**FIGURE 2.1**
Examples of fuzzy set and Boolean set.

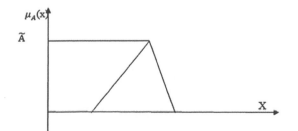

**FIGURE 2.2**
Mapping of a fuzzy set.

A fuzzy set with discrete and finite universe of discontinuous (X) is denoted by

$$\tilde{A} = \left\{ \frac{\mu_{A(x1)}}{x1} + \frac{\mu_{A(x2)}}{x2} + \frac{\mu_{A(x3)}}{x3} + \dots \frac{\mu_{A(xi)}}{xi} \right\}$$

$$= \sum_i \frac{\mu_{A(xi)}}{xi}$$

When the universe of discourse is (X) is continuous and infinite, fuzzy set $\tilde{A}$ is denoted by $\tilde{A} = \int \frac{\mu_{A(xi)}}{xi}$

### 2.4.2 Operation of Fuzzy Sets [21, 22]

#### 2.4.2.1 Union

Let $\tilde{A}$ & $\tilde{B}$ be the two fuzzy sets on the universe X. The union between the sets is denoted by $\mu_{AUB}(x) = \mu_A(x) \mathbf{V} \mu_B(x)$

$$= \max\left( \mu_A(x), \mu_B(x) \right)$$

Let $A = \{(0.1, 1), (0.5, 10), (1, 50)\}$

$$B = \left\{ (1,1), (0.2,10), (0.8,50) \right\}$$

$$A \cap B = \max(A, B)$$

$$= \left\{ (1,1), (0.5,10), (1,50) \right\}$$

#### 2.4.2.2 Intersection

Let $\tilde{A}$ & $\tilde{B}$ be the two fuzzy sets on the universe X. The union between the sets is denoted by

$$\mu_{A \cap B}(x) \quad = \mu_A(x) \wedge \mu_B(x)$$
$$= \min\left( \mu_A(x), \mu_B(x) \right)$$

Let $A = \{(0.1, 1), (0.5, 10), (1, 50)\}$

$$B = \left\{ (1,1), (0.2,10), (0.8,50) \right\}$$

$$A \cup B = \min(A, B)$$

$$= \left\{ (0.1,1), (0.2,10), (0.8,50) \right\}$$

### 2.4.2.3 Complement

The complement of fuzzy set $\tilde{A}$ can be expressed as $\overline{\mu_A(A)} = 1 - \mu(A)$
   Let $A = \{(0.1, 1), (0.5, 10), (1, 50)\}$
   Complement of $A = \bar{A} = \{(0.9, 1), (0.5, 10), (0, 50)\}$

### 2.4.2.4 Difference

The difference of a fuzzy set $\tilde{A}$ with respect to $\tilde{B}$, denoted by $\tilde{A} \mid \tilde{B}$, is the intersection of $\tilde{A}$ and complement of $\tilde{B}$. It can be expressed as

$$\mu_{A|B}(x) = \mu_A(x) \cap \mu_B(x)$$

$$= \min\left(\mu_A(x), \overline{\mu_{B(x)}}\right)$$

### 2.4.3 Properties of Fuzzy Sets

With the exception of the middle rules that are not included, the fuzzy set's attributes are identical to those of crisp sets. As a result of the possibility of overlap between fuzzy sets and their complements, the excluded middle law does not apply to fuzzy sets [23–25].

1. Commutivity:
$$\tilde{A} \cup \tilde{B} = \tilde{B} \cup \tilde{A}$$
$$\tilde{A} \cup \tilde{B} = \tilde{B} \cup \tilde{A}$$

2. Associativity:
$$\tilde{A} \cup (\tilde{B} \cup \tilde{C}) = (\tilde{A} \cup \tilde{B}) \cup \tilde{C}$$
$$\tilde{A} \cap (\tilde{B} \cap \tilde{C}) = (\tilde{A} \cap \tilde{B}) \cap \tilde{C}$$

3. Distributivity:
$$\tilde{A} \cup (\tilde{B} \cap \tilde{C}) = (\tilde{A} \cup \tilde{B}) \cap (\tilde{A} \cup \tilde{C})$$
$$\tilde{A} \cap (\tilde{B} \cup \tilde{C}) = (\tilde{A} \cap \tilde{B}) \cup (\tilde{A} \cap \tilde{C})$$

4. Idempotency
$$\tilde{A} \cup \tilde{A} = \tilde{A}$$
$$\tilde{A} \cap \tilde{A} = \varphi$$

5. Identity
$$\tilde{A} \cup \varphi = \tilde{A} \ \& \ \tilde{A} \cap X = \tilde{A}$$
$$\tilde{A} \cap \varphi = \tilde{A} \ \& \ \tilde{A} \cup X = X$$

6. Transitivity:
$$\text{If } \tilde{A} \le \tilde{B} \le \tilde{C} \text{ then } \tilde{A} \le \tilde{C}$$

   In this case the symbol "≤ "means contained in or equivalent to and "<" means contained in.

7. Involution:
$$\overline{\overline{A}} = A$$

8. Excluded Middle Law:
Does not hold
$$\tilde{A} \cup \overline{\tilde{A}} \ne U$$

   Law of contradiction does not satisfy:
$$\tilde{A} \cap \overline{\tilde{A}} = \varphi$$

9. De Morgan's Law:
$$(\tilde{A} \cap \tilde{B})' = \overline{\tilde{A}} \cup \overline{\tilde{B}}$$
$$(\tilde{A} \cup \tilde{B})' = \overline{\tilde{A}} \cap \overline{\tilde{B}}$$

**PROBLEM**

Let the fuzzy sets $\tilde{A}$ and $\tilde{B}$ be defined as follows:

$$\tilde{A} = \left\{ \frac{1}{2} + \frac{0.5}{3} + \frac{0.3}{4} + \frac{0.2}{5} \right\} \ \& \ \tilde{B} = \left\{ \frac{0.5}{2} + \frac{0.7}{3} + \frac{0.2}{4} + \frac{0.4}{5} \right\}$$

$\pi$ Calculate the union, intersection, complement, and difference

**Solution**

**Union**

$$\tilde{A} \cup \tilde{B} = \max(A, B) = \left\{ \frac{1}{2} + \frac{0.7}{3} + \frac{0.3}{4} + \frac{0.4}{5} \right\}$$

**Intersection**

$$\tilde{A} \cap \tilde{B} = \min(A, B) = \left\{ \frac{0.5}{2} + \frac{0.5}{3} + \frac{0.2}{4} + \frac{0.2}{5} \right\}$$

**Complement**

$$\overline{\tilde{A}} = 1 - \tilde{A} = \left\{ \frac{0}{2} + \frac{0.5}{3} + \frac{0.7}{4} + \frac{0.8}{5} \right\}$$

$$\overline{\tilde{B}} = 1 - \tilde{B} = \left\{ \frac{0.5}{2} + \frac{0.3}{3} + \frac{0.8}{4} + \frac{0.6}{5} \right\}$$

**Difference**

$$\tilde{A} \mid \tilde{B} = \tilde{A} \mid \overline{\tilde{B}} = \min(\tilde{A}, \overline{\tilde{B}}) = \left\{ \frac{0.5}{2} + \frac{0.3}{3} + \frac{0.3}{4} + \frac{0.2}{5} \right\}$$

**PROBLEM**

Consider two fuzzy sets $\tilde{X}$ and $\widetilde{Y}$ given by

$$\tilde{X} = \left\{ \frac{0.9}{4} + \frac{0.6}{2} + \frac{0.5}{7} + \frac{0.8}{9} + \frac{0}{1} \right\} \ \& \ \tilde{Y} = \left\{ \frac{0.1}{4} + \frac{0.5}{2} + \frac{1}{7} + \frac{0.6}{9} + \frac{0.4}{1} \right\}$$

Evaluate: (a) $\tilde{X} \cup \tilde{Y}$, (b) $\tilde{X} \cap \tilde{Y}$, (c) $\overline{\tilde{X}}$, (d) $\overline{\tilde{X} \cup \tilde{Y}}$, (e) $\tilde{X} \cup \overline{\tilde{Y}}$

**Solution**

(a)  $\tilde{X} \cup \tilde{Y} = \max(\tilde{X}, \tilde{Y}) = \left\{ \frac{0.9}{4} + \frac{0.6}{2} + \frac{1}{7} + \frac{0.8}{9} + \frac{0.4}{1} \right\}$

(b)  $\tilde{X} \cap \tilde{Y} = \min(\tilde{X}, \tilde{Y}) = \left\{ \frac{0.1}{4} + \frac{0.5}{2} + \frac{0.5}{7} + \frac{0.6}{9} + \frac{0}{1} \right\}$

(c) $\quad \overline{\tilde{X}} = 1 - \tilde{X} = \left\{ \dfrac{0.1}{4} + \dfrac{0.4}{2} + \dfrac{0.5}{7} + \dfrac{0.2}{9} + \dfrac{1}{1} \right\}$

(d) $\quad \overline{\tilde{X}} \cup \tilde{Y} = \max\left(\overline{\tilde{X}}, \tilde{Y}\right) = \left\{ \dfrac{0.1}{4} + \dfrac{0.5}{2} + \dfrac{1}{7} + \dfrac{0.6}{9} + \dfrac{1}{1} \right\}$

(e) $\quad \tilde{X} \cup \overline{\tilde{Y}} = \max\left(\widetilde{X}, \overline{\tilde{Y}}\right)$

$\qquad \overline{\tilde{Y}} = 1 - \tilde{Y} = \left\{ \dfrac{0.9}{4} + \dfrac{0.5}{2} + \dfrac{0}{7} + \dfrac{0.4}{9} + \dfrac{0.6}{1} \right\}$

Hence, $\tilde{X} \cup \overline{\tilde{Y}} = \max\left(\widetilde{X}, \overline{\tilde{Y}}\right) = \left\{ \dfrac{0.9}{4} + \dfrac{0.6}{2} + \dfrac{0.5}{7} + \dfrac{0.8}{9} + \dfrac{0.6}{1} \right\}$

### 2.4.4 Comparison between Crisp Set or Classical Set and Fuzzy Set

1. In contrast to fuzzy sets, which define values from 0 to 1, crisp sets define values as being between 0 and 1 (i.e., yes for 1 and no for 0).
2. Crisp sets have precise properties; fuzzy sets have imprecise properties.
3. In a crisp set items have a full membership but in fuzzy sets they have a partial membership.
4. The laws of excluded middle and noncontradiction hold for crisp sets, but in the case of fuzzy sets they do not hold.
5. A crisp set is similar to Boolean logic (either 0 or 1), but fuzzy sets capture the degree to which something is true.
6. In a crisp set with crisp boundaries, there is no confusion as to where the set borders are located, while in a fuzzy set with ambiguous bounds, there is uncertainty as to where the set limits are located [26].

### 2.4.5 Composition of Fuzzy Set

1. **Max-Min Composition**
   Two fuzzy relations $R$ and $S$ are defined on sets A, B, and C; then

   $R \le A \times B, S \le B \times C$

   Then composition $R.S = $ Relation from A to C

   $$\mu_T(x,z) = \mu_{R.S}(x,z) = \max\left[\min \mu_s(x,y), \mu_R(y,z)\right]$$

   $$\mu_{R.S} = \wedge_Y \left[\mu_R(y,z) \wedge \mu_S(x,y)\right]$$

## 2. Max–Product Composition

The max product of $R(x, y)$ and $S(y, z)$

$$\mu_T(x,z) = \mu_{R.S}(x,z) = \max\left[\min \mu_R(x,y), \mu_S(y,z)\right]$$

$$\mu_{R.S} = \wedge_Y\left[\mu_R(x,y) \wedge \mu_S(y,z)\right]$$

### 2.4.6 Properties of Fuzzy Composition

1. $\tilde{R}.\tilde{S} \neq \tilde{S}.\tilde{R}$
2. $\left(\tilde{R}.\tilde{S}\right)^{-1} = \tilde{R}^{-1}.\tilde{S}^{-1}$
3. $\left(\tilde{R}.\tilde{S}\right).\tilde{M} = \tilde{R}.\left(\tilde{S}.\tilde{M}\right)$

Equivalence relation: Let the relation R be an equivalence relation if the following these properties are satisfied

1. Reflexibility: $\mu_R(x_i, x_i) = 1$ for $(x_i, x_i) \in R$
2. Symmetry: $\mu_R(x_i, x_j) = \mu_R(x_j, x_i)$ or $\mu_R(x_i, x_i) \in R$ or $(x_i, x_i) \in R$
3. Transitivity: $\mu_R(x_i, x_j) = \lambda_1$ & $\mu_R(x_j, x_k) = \lambda_2$ so $\mu_R(x_i, x_k) = \lambda$

Then $\lambda = \min[\lambda_1, \lambda_2]$.

### 2.4.7 Classical Tolerance Relation

The tolerance relation $R_1$ on universe $X$ is one where only the properties of reflexivity and symmetry are satisfied, otherwise called a proximity relation [22, 25].
$R_1^{n-1} = R_1.R_1.R_1 \ldots R_1 = R$ (equivalence relation)
$[R_1^{n-1} = $ fuzzy tolerance relation$]$

### 2.4.8 Features of Membership Function

A membership function identifies the degree of truthfulness. The membership function for any crisp set can be representing by either 1 or 0. The membership function also is defined by the function whose value remains between 0 and 1 [26].

#### 2.4.8.1 Fuzzy Set

Extension set of classical set is fuzzy set, $A = \{x, \mu_A(x) \mid x \in X\}$
Where $X$ is a universe of discourse and $\mu_A(x)$ is the membership function of $x$

### 2.4.8.2 Features of Fuzzy Sets

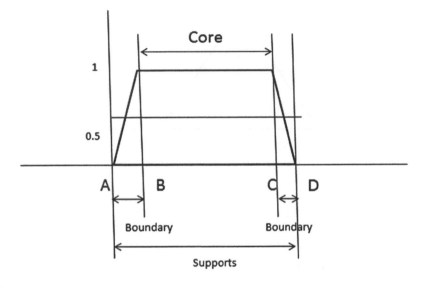

**FIGURE 2.3**
Basic parameters of a membership function [7].

(a) Core: these elements of "$x$" of the universe such that $\mu_A(x) = 1$

(b) Support: These elements of "$x$" of the universe such that $\mu_A(x) \geq 0$

(c) Boundary: Comprise of those elements such that $0 \leq \mu_A(x) \leq 1$

(d) Core = boundary−support

### 2.4.8.3 Classification of Fuzzy Sets

1. **Normal fuzzy set**
   Fuzzy set with at least one "$x$" element whose membership value is unity (Figure 2.4).

2. **Subnormal fuzzy set**
   In this fuzzy set no membership function is equal to unity (Figure 2.5).

3. **Convex fuzzy set**
   In this type of fuzzy set the membership function is monotonically increasing and later monotonically decreasing for elements $x$, $y$, & $z$ (Figure 2.6).

$$\mu_A(x) \geq \min\left[\mu_A(x), \mu_A(z)\right]$$

4. **Non convex fuzzy set**
   It is just opposite to the convex fuzzy set. It is not follows the one particular pattern (monotonically increasing & monotonically decreasing membership function) (Figure 2.7).

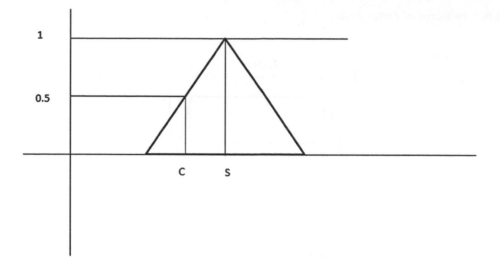

**FIGURE 2.4**
Simple diagram of nominal fuzzy set [7].

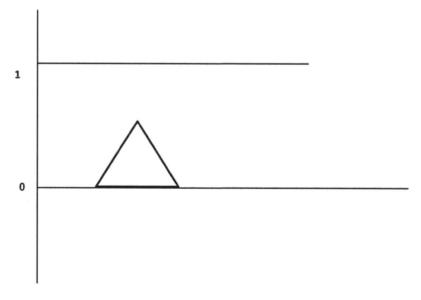

**FIGURE 2.5**
Simple diagram of subnormal fuzzy set [7].

5. **Crossover point**
   When membership value for any value of "$x$" is 0.5 then it is called crossover point membership function (Figure 2.8).

6. **Height of a fuzzy set**
   Maximum value of the membership function for any values of $x$ is the known height of a fuzzy set, where Height $\mu_A(x) = \max\left[\mu_A(x)\right]$ (Figure 2.9).

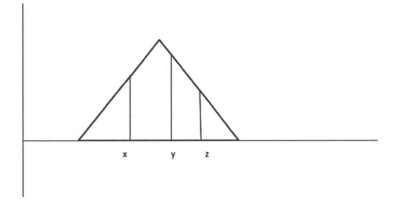

**FIGURE 2.6**
Simple diagram of convex fuzzy set [7].

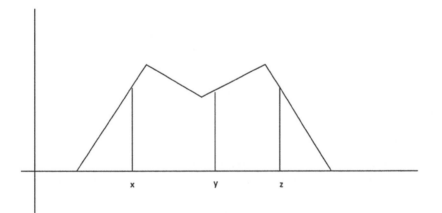

**FIGURE 2.7**
Simple diagram of non convex fuzzy set [7].

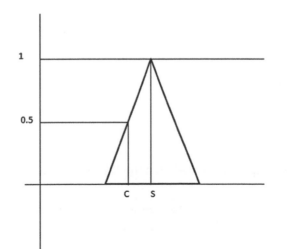

**FIGURE 2.8**
Simple diagram of crossover fuzzy set [7].

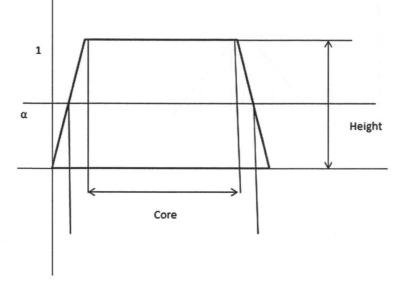

**FIGURE 2.9**
Simple diagram of height of a fuzzy set [7].

## 2.5 Fuzzification

Fuzzification is the procedure used to change a crisp value into a fuzzy value or a crisp value (existing value) into a linguistic value [27–29]. Take an example where we want to convert the crisp temperature 10°C into a fuzzy value (Figure 2.10)
Methods of membership value assignment:

1. Intitution
2. Inference
3. Rank ordering
4. Angular fuzzy sets
5. Neural network
6. Genetic algorithm
7. Inductive reasoning

### 2.5.1 Institution

Design the membership by humans' own intelligence (Figure 2.11).

**FIGURE 2.10**
Simple diagram of fuzzification [7].

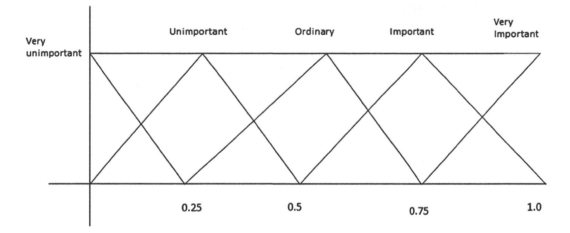

**FIGURE 2.11**
Simple diagram of fuzzy institution [8].

### 2.5.2 Inference

Inference uses knowledge to perform reasoning. Knowledge of geometrical shapes membership values includes triangular, trapezoidal, bell shaped, and Gaussian as several common membership functions. Let A, B, C be the interior angle of triangle $A \geq B \geq C > 0$ and $A + B + C = 180°$.

Here we are defining five types of triangles:

1. R = Approximately right angle triangle
2. I = Approximately isosceles triangle
3. E = Approximately equilateral triangle
4. I.R = Approximately isosceles and right angle triangle
5. T = Other types of triangle

1. Membership value of right angle triangles:

$$\mu_A(A,B,C) = 1 - \frac{1}{90}|A - 90|$$

Example: $\mu = \{80, 65, 35\}$

$$\mu_R(A,B,C) = 1 - \frac{1}{90}|A - 90|$$

$$= 1 - \frac{1}{90}|80 - 90| = 1 - \frac{1}{9} = \frac{8}{9}$$

2. Membership values of isosceles triangles:

$$\mu_I(A,B,C) = 1 - \frac{1}{60}\min\{(A-B),(B-C)\}$$

Let, $\mu(A,B,C) = \{80,65,35\}$

Then $\mu_I(A,B,C) = 1 - \dfrac{1}{60}\min\{(80-65),(65-30)\}$

$$= 1 - \dfrac{1}{60}\min\{15,30\}$$

$$\mu_I(A,B,C) = 1 - \dfrac{1}{60} * 15 = \dfrac{3}{4}$$

3. Membership value of equivalent triangles:

$$\mu_E(A,B,C) = 1 - \dfrac{1}{180}|A - C|$$

Let, $\mu(A, B, C) = \{80, 65, 35\}$

$$\mu_E(A,B,C) = 1 - \dfrac{1}{180}|80 - 35|$$

$$= 1 - \dfrac{1}{180} * 45 = 1 - \dfrac{1}{4} = \dfrac{3}{4}$$

4. Membership value of isosceles and right angle triangles:

$$IR = I \cap R$$

$$\mu_{IR}(A,B,C) = \min\{\mu_I(A,B,C),\mu_R(A,B,C)\}\}$$

Let, $\mu(A, B, C) = \{80, 65, 35\}$

$$\mu_{IR}(A,B,C) = \min\{\mu_I,\mu_R\}$$

$$= \min\left(\dfrac{3}{4},\dfrac{8}{9}\right) = \dfrac{3}{4}$$

5. Membership value of other types of triangle:

$$T = (R \cup I \cup E)^c$$

$$T = R^c \cap I^c \cap E^c$$

Let, $\mu(A, B, C) = \{80, 65, 35\}$;

$$\mu_R = \dfrac{8}{9}; \mu_I = \dfrac{3}{4} \;\&\; \mu_E = \dfrac{3}{4}$$

$$\mu_R^c = 1 - \frac{8}{9} = \frac{1}{9}; \ \mu_I^c = 1 - \frac{3}{4} = \frac{1}{4}; \ \mu_E^c = 1 - \frac{3}{4} = \frac{1}{4}$$

$$\mu_T = R^c \cap I^c \cap E^c = \frac{1}{9} \cap \frac{1}{4} \cap \frac{1}{4} = \frac{1}{4}$$

### 2.5.3 Rank Ordering

By using a rank ordering procedure, the polling notion is employed to assign membership value (pairwise preference). From the ordering, membership is performed.

| Name of the Car | BMW | Benz | Jaguar | Audi | Total | % |
|---|---|---|---|---|---|---|
| BMW | — | 51 | 54 | 52 | 157 | 26.03 |
| Benz | 48 | | 47 | 84 | 179 | 32.66 |
| Jaguar | 46 | 62 | | 14 | 122 | 20.33 |
| Audi | 45 | 53 | 47 | | 145 | 24.04 |

Based on the percentage, membership values are assigned (Figure 2.12).

### 2.5.4 Angular Fuzzy Sets

These are different from standard fuzzy sets in this co-ordinate description: They depend upon the universe of angles repeating the shapes every $2\pi$.

Example: pH value of water sample from contaminated pond.

a. If pH is 7, it is a neutral solution.
b. A pH between 7 to 14 means absolute basic (AB) $\left(\frac{\pi}{2}\right)$, very basic (VB) $\left(\frac{3\pi}{8}\right)$, basic (B) $\left(\frac{\pi}{4}\right)$, and medium basic (MB) $\left(\frac{\pi}{8}\right)$.
c. If pH is between 7 and neutral (0), it is medium acidic (MA) $\left(-\frac{\pi}{8}\right)$, very acidic (VA) $\left(-\frac{3\pi}{8}\right)$, or acidic (A) $\left(-\frac{\pi}{4}\right)$.
d. Membership function $M_t(\theta) = t \tan(\theta)$.

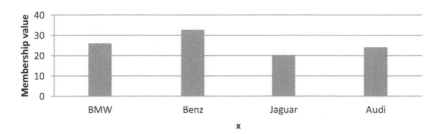

**FIGURE 2.12**
Simple diagram of rank ordering membership function [8].

### 2.5.5 Neural Network

Here, $R_A$, $R_B$ & $R_C$ are the data classifier in region [like red, yellow, and black balls]. $x$, $y$ are the data points (data coordination) (Figure 2.13).

From the diagram it is shown that data points lie in the $R_C$. 100 data points lies in $R_A$ and so forth.

### 2.5.5.1 A Training the Neural Network

All the data points of $(x_1, x_2)$ training and all possible combinations of $R_A$, $R_B$, and $R_C$ are stored in the neural network (Figure 2.14).

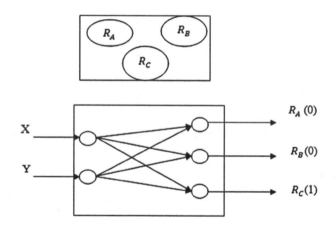

**FIGURE 2.13**
Simple diagram of ANN model [8].

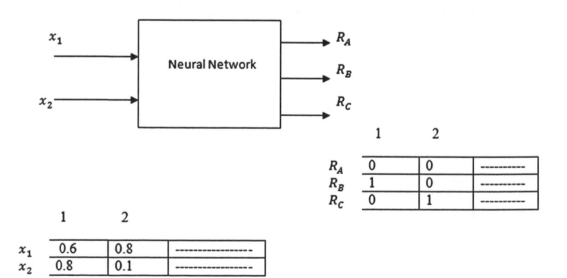

|       | 1 | 2 |             |
|-------|---|---|-------------|
| $R_A$ | 0 | 0 | ----------- |
| $R_B$ | 1 | 0 | ----------- |
| $R_C$ | 0 | 1 | ----------- |

|       | 1   | 2   |                    |
|-------|-----|-----|--------------------|
| $x_1$ | 0.6 | 0.8 | ------------------ |
| $x_2$ | 0.8 | 0.1 | ------------------ |

**FIGURE 2.14**
Simple diagrams for training the neural network model [8].

### 2.5.5.2 Testing the Neural Network

When new data points come (which are not used to train the model) the neural network compares it with the train model from the available database to give the approximate output 30–33 (Figure 2.15).

### 2.5.6 Genetic Algorithm

GA used to determine the fuzzy membership function. Base on Darwin's idea of evaluation of "survival of the fittest." Procedure and methods [30]:

1. The membership function are called into bit strings.
2. The bit strings are calculated together.
3. The fitness function is used to evaluate each set of membership functions.
4. The process is carried out until convergence is achieved.

### 2.5.7 Inductive Reasoning

Logical thinking, observation, and experiment sum up to indicate the conclusion of the membership function. As an example, during sunny weather, you don't need an umbrella.

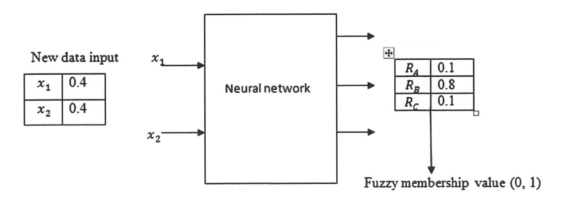

**FIGURE 2.15**
Simple diagrams for testing the neural network model [8].

## 2.6 Defuzzification

**FIGURE 2.16**
Simply diagram of defuzzification [8].

### 2.6.1 Lambda Cut for Fuzzy Sets/Alpha Cut

Defuzzification is the process where we can convert the fuzzy value into the crisp value or fuzzy set into crisp value or fuzzy matrix into a single value [30–32].

Defuzzification can be done by the lambda cut method or alpha cut method.

Condition

$$A_\lambda = \{x \mid \mu_A(x) \geq \lambda\}; \lambda \epsilon(0,1)$$

a) Strong $\lambda$ cut defuzzification: When $\mu_A(x) > \lambda$
b) Weak $\lambda$ cut defuzzification: When $\mu_A(x) \geq \lambda$

Properties of $\lambda$ cut set

$$(1) \left( \acute{A} \cup \acute{B} \right)_\lambda = A_\lambda \cup B_\lambda$$

$$(2) \left( \acute{A} \cap \acute{B} \right)_\lambda = A_\lambda \cap B_\lambda$$

$$(3) \left( \overline{\overset{\vee}{A}} \right)_\lambda \neq \left( \overline{A} \right)_\lambda \left[ \text{except} \lambda = 0.5 \right]$$

Lambda cut for fuzzy relation

Let $\overset{\vee}{R}$ be a fuzzy relation

$$R_\lambda = \left\{ (x,y) \mid \mu_R(x,y) \geq \lambda \right\}$$

For two fuzzy relations $\overset{\vee}{R}$ and $\overset{\vee}{S}$

$$(1) \left( \acute{R} \cup \acute{S} \right)_\lambda = R_\lambda \cup S_\lambda$$

$$(2) \left( \acute{R} \cap \acute{S} \right)_\lambda = R_\lambda \cap S_\lambda$$

$$(3) \left( \overline{\overset{\vee}{R}} \right)_\lambda \neq \left( \overline{S} \right)_\lambda [ \text{except } \lambda = 0.5]$$

**PROBLEM**

Consider a fuzzy set

$$\overset{\vee}{A} = \left\{ \frac{0.4}{x_1} + \frac{0.6}{x_2} + \frac{0.5}{x_3} + \frac{0.7}{x_4} \right\} \& A_{0.5} = \left\{ x_1, x_2, x_3, x_4 \right\}$$

Find the crisp set value membership value whose value is greater than or equal to 0.5.

**Solution:**

$$A_{0.5} = \left\{ x_2, x_3, x_4 \right\}$$

$$A_{0.6} = \left\{ x_2, x_4 \right\}$$

$$A_{0.7} = \left\{ x_4 \right\}$$

Defuzzification method [5, 28, 29, 33]

1. Max membership method
2. Centroid method
3. Weighted average method
4. Mean–max membership
5. Center of sums
6. Center of largest area
7. First of maxima, last of maxima

## 2.6.2 Max Membership Principle

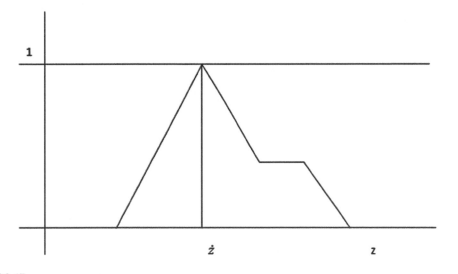

**FIGURE 2.17**
Simple diagram of max membership method [8].

a. This is the height method
b. Maximum membership value
c. $z^*$ = defuzzified value or crisp value
d. $\mu_A(z^*) \gg \mu_A(z)$ for all $z \epsilon A$
e. Peaked output function

There are three membership function 0.3, 0.5, and 1 and $z^* = 1$ is the maximum crisp value [34].

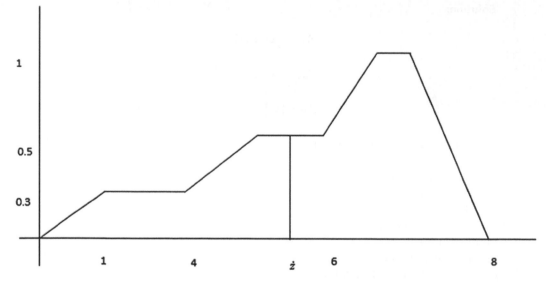

**FIGURE 2.18**
Max membership defuzzification methods [9].

### 2.6.3 Centroid Method

Centroid membership function means center of mass, center of area, or center of gravity [35].

$$z^* = \frac{\int \mu_A . z\, dz}{\int \mu_A . dz}$$

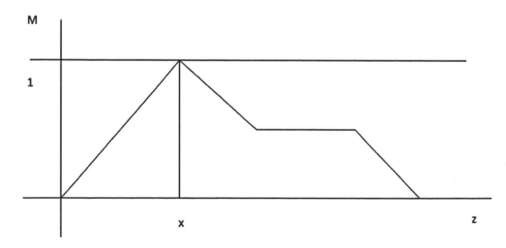

**FIGURE 2.19**
Centroid membership defuzzification methods [9].

### 2.6.4 Weighted Average Method

This is true for symmetric output membership functions. Each membership function is weighed based on its maximum membership value [30].

$$z^* = \frac{\sum \mu_{A(\bar{z})} \bar{z}}{\sum \mu_A (\bar{z})}$$

Let there be three membership functions with center values *a*, *b*, and *c*. Then membership value

$$z^* = \frac{(0.3 * a + 0.5 * b + 1 * * c)}{(0.3 + 0.5 + 1)}$$

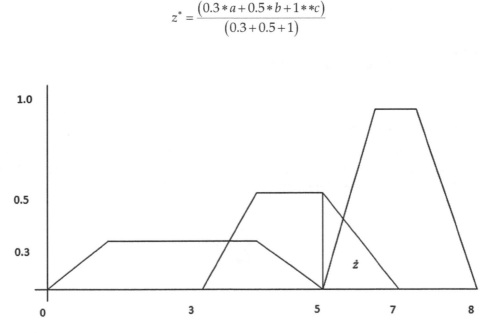

**FIGURE 2.20**
Weighted membership defuzzification method [9].

### 2.6.5 Mean–Max Membership or Middle of Maxima

The center of the maximum is another name for this approach [32]. With the exception of the possibility of non-unique maximum membership regions, this is closely comparable to the max membership approach. The output of the resulting membership function is given by

$$z^* = \frac{\sum z_i}{n} = \frac{a+b}{2}$$

This type of membership function is only applicable for a symmetrical membership function.

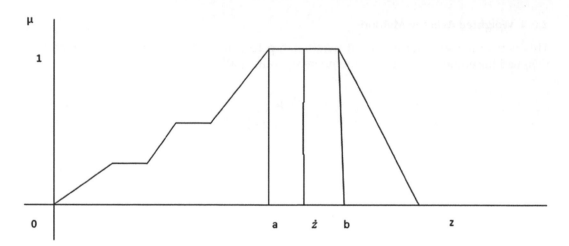

**FIGURE 2.21**
Mean max membership defuzzification method [9].

### 2.6.6 Center of Sum Methods

Instead of using their union, this approach uses the algebraic sum of each fuzzy subset. The crossing regions are added twice, which is the biggest disadvantage despite the fact that the computations are done quite quickly. The defuzzified value $z^*$ is given by

$$z^* = \frac{\sum_{i=1}^{n} z_i \, A_{ci}}{\sum_{i=1}^{n} A_{ci}}$$

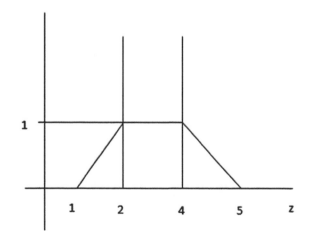

**FIGURE 2.22**
Center of sum membership defuzzification method [9].

For fuzzy set 1:

$$z_1 = \frac{(1+5)}{2} = 3,$$

$$A_1 = \frac{(2+4)*0.2}{2} = 0.6$$

Fuzzy set 2:

$$z_2 = \frac{(3+7)}{2} = 5,$$

$$A_2 = \frac{(2+4)*1}{2} = 3$$

$$z^* = \frac{(0.6*3+3*5)}{(0.6+3)} = 4.666$$

### 2.6.7 Center of Largest Area

If the fuzzy set comprises several subregions, the center of biggest area approach may be used to identify the subregion with the largest area.

$$z^* = \frac{\int \mu_c.z\,dz}{\int \mu_c.dz} = \frac{(11.5+12.5)}{2} = 12$$

**Problem** Calculate the different defuzzification membership values for the following Figure 2.24.

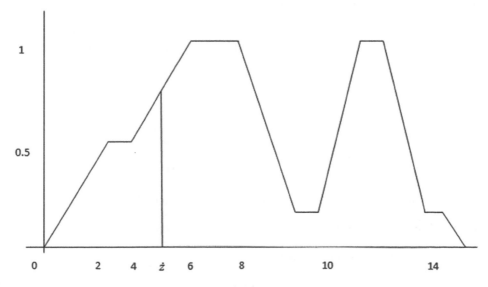

**FIGURE 2.23**
Center of largest area of membership defuzzification method [9].

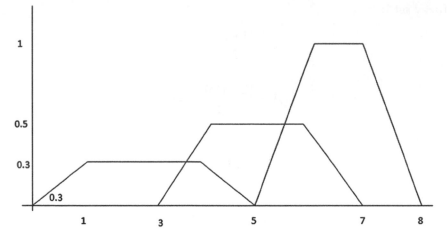

**FIGURE 2.24**
Union of three fuzzy set diagrams [9].

## 2.7 Examples for Different Defuzzification Methods

### 2.7.1 Max Membership Method

From the diagram it is seen that the maximum membership value is 1, and the equivalent scalar value is 1 where $z^* = 6, 7$

### 2.7.2 Centroid Method

Finding the region limited by the union of the three sets and determining its centroid—which will serve as the defuzzified value.

Sub area 1: Area $= \dfrac{1 * 0.3}{2} = 0.150,$      $\bar{z} = 0.67,$      Area $* \bar{z} = 0.100$

Sub area 2: Area $= 3 * 0.3 = 0.90$      $\bar{z} = 2.5$      Area $* \bar{z} = 2.250$

Sub area 3: Area $= \dfrac{0.4 * 0.2}{2} = 0.04$      $\bar{z} = 3.73$      Area $* \bar{z} = 0.149$

Sub area 4: Area $= 2 * 0.5 = 1.00$      $\bar{z} = 1.00$      Area $* \bar{z} = 5.00$

Sub area 5: Area $= \dfrac{0.5 * 0.5}{2} = 0.125$      $\bar{z} = 5.87$      Area $* \bar{z} = 7.330$

Sub area 6: Area $= 1 * 1 = 1$      $\bar{z} = 6.50$      Area $* \bar{z} = 6.50$

Sub area 7: Area $= \dfrac{1 * 1}{2} = 0.50$      $\bar{z} = 7.33$      Area $* \bar{z} = 3.660$

Total area $= 3.715$ & $\sum \bar{z}$. Area $= 24.989$

$$z^* = \frac{24.989}{3.715} = 6.720$$

### 2.7.3 Weighted Average Method

In this instance, the mean is multiplied by the degree of membership of each fuzzy set to get the Centre, and the average of all sets is then determined.

The membership value of first fuzzy (trapezoidal diagram) at 2.5 is 0.3. The membership value of second fuzzy (trapezoidal diagram) at 0.5 is 0.5 & the membership value of first fuzzy (trapezoidal diagram) at 6.5 is 1

$$\text{So, } z^* = \frac{(2.5*0.3+0.5*1+6.5*1)}{(0.3+0.5+1)} = 5.146$$

### 2.7.4 Mean Max Membership

Here, the average of the range with the highest membership value. The maximum membership value in this case is 1 for the range (6, 7). Therefore $z^* = \dfrac{6+7}{2} = 6.5$

### 2.7.5 Center of Sums

In this approach, we first identify the centre of the region by adding the areas of each fuzzy set.

$$\text{Area of the first fuzzy set} = \frac{(5+3)*0.3}{2} = 1.2$$

$$\text{Area of the second fuzzy set} = \frac{(4+2)*0.5}{2} = 1.5$$

$$\text{Area of the third fuzzy set} = \frac{(3+1)*1}{2} = 2$$

$$\text{So, } z^* = \frac{\left[(1.2*2.5)+(1.5*5)+(2*6.5)\right]}{(1.2+1.5+2)} = 5$$

### 2.7.6 Center of Largest Area

According to this procedure, the mean value of the fuzzy set with the greatest or maximum area is called the defuzzified value. From the diagram it is seen that the third fuzzy set has the largest area. Therefore $z^* = \dfrac{(6+7)}{2} = 6.5$.

## References

1. Teodorović D. Fuzzy logic systems for transportation engineering: The state of the art. *Transportation Research Part A: Policy and Practice.*
2. Ross TJ. *Fuzzy logic with engineering applications.* John Wiley & Sons;2005.
3. McNeill FM, Thro E. *Fuzzy logic: A practical approach.* Academic Press;2014.
4. Mukaidono M. *Fuzzy logic for beginners.* World Scientific;2001.
5. Hasuike T, Katagiri H, Ishii H. Portfolio selection problems with random fuzzy variable returns. *Fuzzy Sets and Systems* 2009;160(18): 2579–2596.

6. Kasabov NK. *Foundations of neural networks, fuzzy systems, and knowledge engineering.* Marcel Alencar;1996.

7. Friedlob GT, Schleifer LL. *Fuzzy logic: application for audit risk and uncertainty. Managerial Auditing Journal* 2009;14(3): 127–137.

8. Rihoux B, Grimm H. Diversity, ideal types and fuzzy sets in comparative welfare state research. In *Innovative comparative methods for policy analysis,* Springer;2006, pp. 167–184.

9. McNeill D, Freiberger P. *Fuzzy logic: The revolutionary computer technology that is changing our world.* Simon and Schuster;1994.

10. Pedrycz W, Gomide F. *An introduction to fuzzy sets: Analysis and design.* MIT Press;1998.

11. Badiru AB, Cheung J. *Fuzzy engineering expert systems with neural network applications.* John Wiley & Sons;2002, Vol. 11.

12. Bandarian R. Measuring commercial potential of a new technology at the early stage of development with fuzzy logic. *Journal of Technology Management & Innovation* 2007;2(4): 73–85.

13. Ragin CC. *Fuzzy-set social science.* University of Chicago Press;2000.

14. Kovacic Z, Bogdan S. *Fuzzy controller design: Theory and applications.* CRC Press;2018.

15. Wang H, Xu Z, Pedrycz W. An overview on the roles of fuzzy set techniques in big data processing: Trends, challenges and opportunities. *Knowledge-Based Systems* 2017;118: 15–30.

16. Chow MY. *Methodologies of using neural network and fuzzy logic technologies for motor incipient fault detection.* World Scientific;1997.

17. Dutta P, Kumar A. Intelligent calibration technique using optimized fuzzy logic controller for ultrasonic flow sensor. *Mathematical Modelling of Engineering Problems* 2017;4(2): 91–94.

18. Dutta P, Kumar A. Design an intelligent flow measurement technique by optimized fuzzy logic controller. *Journal Europen des Systmes Automatiss* 2018;51: 89–107.

19. Plewa C, Ho J, Conduit J, Karpen IO. Reputation in higher education: A fuzzy set analysis of resource configurations. *Journal of Business Research* 2016;69(8): 3087–3095.

20. Kosko B. Fuzziness vs. Probability. *International Journal of General System* 1990;17(2–3): 211–240.

21. Klir G, Yuan B. *Fuzzy sets and fuzzy logic.* Prentice Hall;1995, Vol. 4, pp. 1–12.

22. Adeli H, Sarma KC. *Cost optimization of structures: Fuzzy logic, genetic algorithms, and parallel computing.* John Wiley & Sons;2006.

23. Terano T, Asai K, Sugeno M. (Eds.). *Applied fuzzy systems.* Academic Press;2014.

24. Alonso JM, Magdalena L, González-Rodríguez G. Looking for a good fuzzy system interpretability index: An experimental approach. *International Journal of Approximate Reasoning* 2009;51(1): 115–134.

25. Adriaenssens V, De Baets B, Goethals PL, De Pauw N. Fuzzy rule-based models for decision support in ecosystem management. *Science of the Total Environment* 2004;319(1–3): 1–12.

26. Bělohlávek R, Dauben JW, Klir GJ. *Fuzzy logic and mathematics: A historical perspective.* Oxford University Press;2017.

27. Papageorgiou EI, Salmeron JL. A review of fuzzy cognitive maps research during the last decade. *IEEE Transactions on Fuzzy Systems* 2012;21(1): 66–79.

28. Pedrycz W, Gomide F. *Fuzzy systems engineering: Toward human-centric computing,* John Wiley & Sons; 2007.

29. Aliev RA. *Fundamentals of the fuzzy logic-based generalized theory of decisions.* Springer;2013.

30. Sen Z. *Fuzzy logic and hydrological modeling.* CRC Press;2009.

31. Passino KM, Yurkovich S, Reinfrank M. *Fuzzy control.* Addison-wesley;1998, Vol. 42, pp. 15–21.

32. Hannan MA, Ghani ZA, Hoque MM, Ker PJ, Hussain A, Mohamed A. Fuzzy logic inverter controller in photovoltaic applications: Issues and recommendations. *IEEE Access* 2019;7: 24934–24955.

33. Benitez JM, Martín JC, Román C. Using fuzzy number for measuring quality of service in the hotel industry. *Tourism Management* 2007;28(2): 544–555.

34. Zimmermann HJ. *Fuzzy sets, decision making, and expert systems.* Springer Science & Business Media;1987, Vol. 10.

35. Momoh JA, Ma XW, Tomsovic K. Overview and literature survey of fuzzy set theory in power systems. *IEEE Transactions on Power Systems* 1995;10(3): 1676–1690.

# 3

## A Practical Approach to Neural Network Models

### 3.1 Introduction

An effective and parallel distributed computing system called ANN borrows the analogy of biological neural networks for its main idea [1, 2]. It acquires a significant collection of units that are connected in some way in order to facilitate inter-unit communication. These parts, sometimes called nodes or neurons, are simple parallel processors [3]. Each neuron is connected to other neurons via a connection. A weight with knowledge of the input signal is connected to each connecting link. Since the weight often stimulates or inhibits the information being transmitted, this knowledge is particularly useful for helping neurons solve a particular problem [4]. The intrinsic state of each neuron is described by an activation signal [5]. A combination of the input signals and the activation rule results in output signals that may be delivered to other components (Figure 3.1).

The net input for the aforementioned generic artificial neural network model may be approximated:

$$Y_{in} = w_1 x_1 + w_2 x_2 + w_3 x_3 + \ldots + w_n x_n$$

That's the net input $Y_{in} = \sum_{i=1}^{n} x_i w_i$

Applying the activation function to the net input enables the determination of the output.

$$Y = f\left(Y_{in}\right)$$

Output = function (net input calculated)

The three building pieces that make up ANN processing are as follows:

- Network topography
- Weight or learning adjustments
- Activation techniques

#### 3.1.1 Network Topology

This is an arrangement of a network which indicates how many different ways nodes or connecting lines can be arranged. According to the network topology, ANN can be classified into main two categories as shown in Figure 3.2.

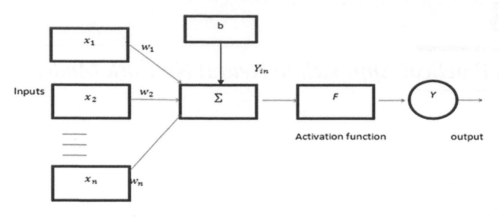

**FIGURE 3.1**
Typical architecture of ANN [3].

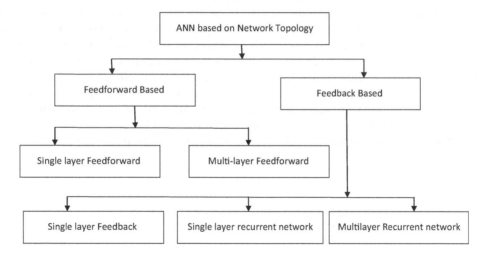

**FIGURE 3.2**
Types of ANN model based on connecting nodes.

### 3.1.1.1 Feed Forward Network

The term "feed forward layer" refers to a non-recurrent network containing layers of processing units and nodes, each of which is connected to the nodes of the previous layers [5]. There are various weights attached to the connection. The signal can be unidirectional, from input to output, because there is no feedback loop. It may be divided into the following two types: multilayer feed forward and single layer feed forward network.

**Single layer feed forward network**
A single layer feed forward ANN has only a single weighted layer, or the input layer and output layer are both fully interconnected, as shown in Figure 3.3.

**Multilayer feed forward network**
In a multilayer feed forward ANN architecture there are multiple weighted layers, and between the input and output layers there will be one or more hidden layers available as shown in Figure 3.4.

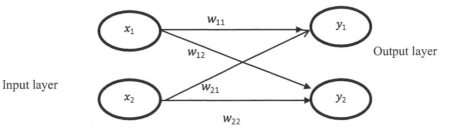

FIGURE 3.3
Basic diagrams of single layer feed forward network.

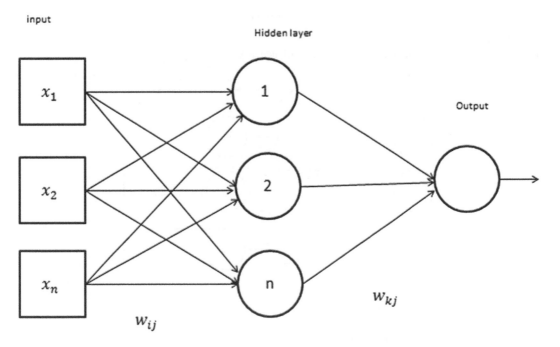

FIGURE. 3.4
Basic diagrams of multilayer feed forward network [5].

### 3.1.1.2 Feedback Network

A feedback network, as its title implies, contains feedback pathways, permitting the signal to go through loops in both directions. As a result, it is a nonlinear dynamic system that evolves continually until it achieves equilibrium [6, 7]. Unlike feed forward, a feedback network may be classified into the following categories.

#### Single node with its own feedback
When outputs may be utilized as inputs for nodes in the same layer or a layer above, single node feedback ANN feedback networks arise. Recurrent networks are closed-loop feedback networks [6]. Figure 3.5 represents a single neuron in a single recurrent network that receives feedback from another neuron.

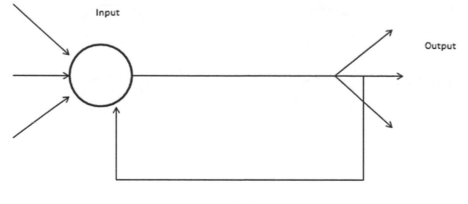

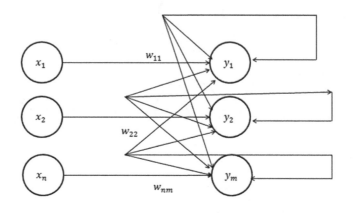

**FIGURE 3.5**
Basic diagrams of single nodes with their own feedback network.

**FIGURE 3.6**
Basic diagrams of single layer recurrent networks.

### Single layer recurrent network

A feedback link is formed in a single layer ANN in which a processing unit's output can be directed back to that processing element, to another processing element, or to both, as illustrated in Figure 3.6. An ANN belonging to the class of recurrent neural networks has connections between nodes that form directed graphs along a sequence [6–8]. It may therefore show dynamic temporal behavior for a time series. RNNs may use their memory to process input sequences, unlike a feed forward ANN.

### Multilayer recurrent network

A multilayer recurrent network has an internal, hidden layer that is not in direct communication with the exterior layer as shown in Figure 3.7. The network becomes more computationally robust when one or more hidden layers are present. The model's outputs cannot be fed back into it because there are no feedback connections. In every step it does not need to calculate the value of the input. The main advantage of RNN is that the hidden layer captures the information in a sequential way.

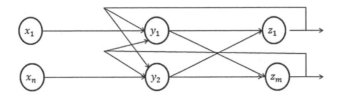

**FIGURE 3.7**
Basic diagram of a multilayer recurrent network.

## 3.1.2 Adjustments of Weights or Learning

Learning is an important parameter in artificial neural network which helps to modify the weights of the connection made inside a particular network's neurons. Based on learning methods, it may be divided into three groups in the ANN, namely supervised learning, unsupervised learning, and reinforcement learning as shown in Figure 3.8.

### 3.1.2.1 Supervised Learning

This form of algorithm for learning is carried out under the supervision of a teacher, as the name suggests. It depends on this learning method [9–11]. When an ANN is being trained using supervised learning, the input vector is sent to the network, and the network will produce an output vector as shown in Figure 3.9. The desired output vector is contrasted

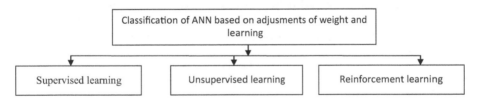

**FIGURE 3.8**
Classification of ANN model based on learning and weight.

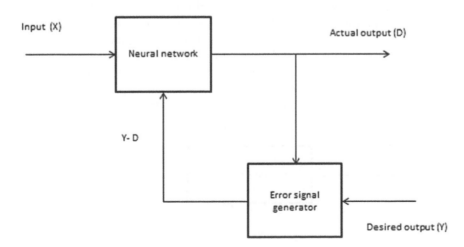

**FIGURE 3.9**
Basic diagrams of supervised learning.

with this one. An erroneous signal is produced if the planned output vector and the corresponding data differ. For as long as the expected output is not yet met by the actual output, the weights are adjusted depending on this error signal. Figure 3.9 represents the simple diagram of supervised learning.

### 3.1.2.2  Unsupervised Learning

Unsupervised learning enables the learning of datasets without the constant monitoring of a professional [12, 13]; the whole process is autonomous and is depicted in Figure 3.10. By mixing input vectors of the same kind, clusters are produced when ANNs are being trained via unsupervised learning. The neural network interacts to a new input pattern by identifying the class to which the pattern belongs in a simulation result. Figure 3.10 illustrates that in unsupervised learning the environment doesn't give feedback on whether the planned result was achieved or not. As a result, in this sort of learning, the network itself updates its weighted value and identifies the patterns [14, 15], characteristics, and relationships between the input and output data.

### 3.1.2.3  Reinforcement Learning

In reinforcement learning ANN, a learning algorithm is employed to enhance or reinforce the network in relation to some critical knowledge [16, 17]. The learning topology of this ANN is quite similar to the supervised learning. The network gets some feedback from the environment as it is being trained using reinforcement learning as shown in Figure 3.11. A feedback network provides numerical information instead of instructive information. The network adjusts the weights after obtaining input in order to receive better criticism in the future [18, 19].

### 3.1.3  Activation Functions

The activation function chooses whether or not to stimulate a neuron by calculating a weighted sum and adding bias to it. Adding nonlinearity to a neuron's output is the goal of the activation function. We know that the weight, bias, and matching activation functions of neurons in neural networks determine how well they function. The output inaccuracy would be used for a neural network's neurons' weights and biases that can be modified. This process's official term is back-propagation. Back-propagation is made possible by activation functions since they provide the gradients and error needed to update the weights and biases.

The following list includes several interesting activation functions–

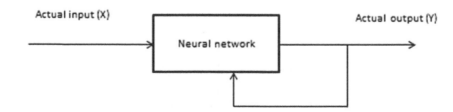

**FIGURE 3.10**
Diagram of unsupervised learning.

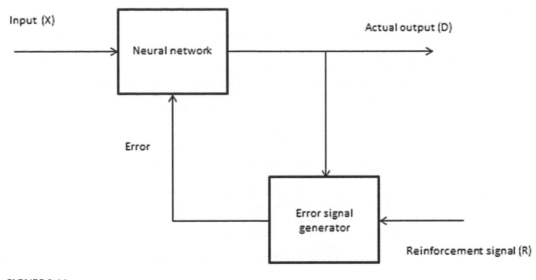

**FIGURE 3.11**
Diagram of reinforcement learning.

### 3.1.3.1 Type of Activation Function

**TABLE 3.1**

Characteristics of Different Types of Activation Function [20–22]

| Si No. | Name of the Activation Function | Characteristics |
|---|---|---|
| 1 | **Linear activation function:**<br>Linear activation function is similar to a straight line passing through the equation $y = mx$, which is identical to that of a straight line for an origin with a positive slope, as illustrated in Figure 3.12. Irrespective of the number of layers, assuming they are all linear in nature, the last layer's final activation function is little more than a straight function that can be entered from the first layer. Its values range from $-\alpha$ to $+\alpha$. It is utilized at the output layer. Differentiating a linear function will cause the function to become constant and the outcome to no longer be dependent on the input "x." | See Figure 3.12 |
| 2 | **Sigmoid activation function**<br>This activation function looks like an 'S' shaped graph and is mathematically represented by $f(x) = \dfrac{1}{1+e^x}$. From the characteristics graph it is seen that the function is nonlinear in nature, and output values are very steep corresponding to $x$ axis as shown in Figure 3.13. The magnitude of the $y$ axis ranges from 0 to 1. Output of this function can easily predict 1 when its value exceeds 0 and 0.5. | See Figure 3.13 |
| 3 | **Tanh activation function**<br>Tanh activation function is sometimes also called the "always works better than sigmoid function" or the tangent hyperbolic function as shown in Figure 3.14. Actually, it's a shifted nonlinear variant of the sigmoid function. Both are comparable and derivable from one another. Mathematically it is represented by $f(x) = \tanh(x) = \dfrac{2}{1+e^{-2x}} -1$. Output of this activation function ranges from $-1$ to $+1$. This greatly simplifies learning for the subsequent layer. | See Figure 3.14 |

(*Continued*)

**TABLE 3.1 (*Continued*)**

| Si No. | Name of the Activation Function | Characteristics |
| --- | --- | --- |
| 4 | **ReLu activation function** <br> The rectified linear unit is referred to as ReLu. It is the activation function that is utilized the most, and is mostly used in the underlying layers of neural networks as shown in Figure 3.15. This nonlinear mathematical function can be represented by $A(x) = \max(0, x)$ and its output varies from $[0, a]$. Because ReLu uses fewer complicated mathematical processes than tanh and sigmoid, it requires less computing power. When just a few neurons are active at once, the network is sparse, which makes computation simple and effective. Compared to sigmoid and tanh functions, the ReLu activation function learns a lot more effectively. | See Figure 3.15 |
| 5 | **Softmax function** <br> The Softmax activation function is one kind of nonlinear sigmoid function which is quite handy during the classification problems shown in Figure 3.16. Its output squeeze for each class is between 0 and 1. The Softmax function is applicable at the classifier's output layer, which is where we are truly aiming to get the probabilities to categorize each input. | See Figure 3.16 |

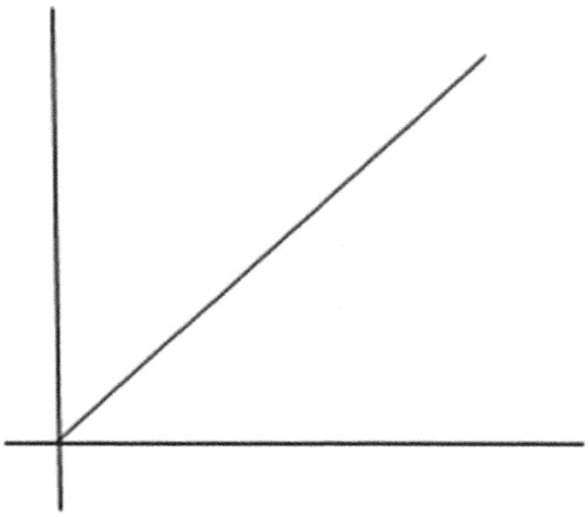

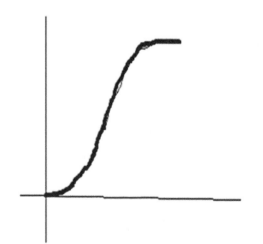

**FIGURE 3.12**
Characteristics graph for linear function.

**FIGURE 3.13**
Characteristics graph for sigmoid function.

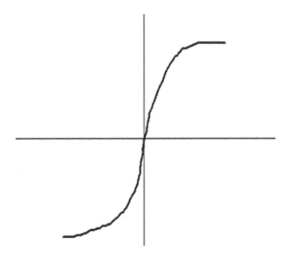

**FIGURE 3.14**
Characteristics graph of tanh activation function.

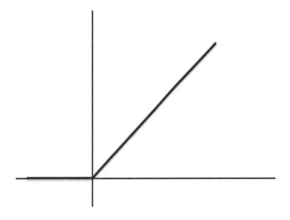

**FIGURE 3.15**
Characteristics graph for ReLu activation function.

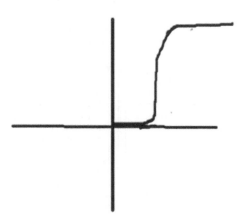

**FIGURE 3.16**
Characteristics graph for Softmax activation function

### 3.1.4 Learning Rules in Neural Network

In order to increase the performance of a neural network model, learning rules, also known as the learning process, enable updating the network's weights and bias levels while emulating a particular data pattern. During the simulation a neural network applying this iterative process the learning rule is an iterative process [23, 24]. A neural network can improve its operation better by learning from the current circumstances [25]. Let's examine various neural network learning rules as shown in Figure 3.17.

#### 3.1.4.1 Hebbian Learning Rule

The Hebbian rule was the first learning rule, created in 1949 by Donald Hebb as a learning technique for an unsupervised neural network. It can help us figure out how to increase a network's node weights [26, 27]. The Hebb learning rule makes the assumption that two neighboring neurons must have simultaneous activation and deactivation. The weight linking these neurons should consequently get heavier. The weight between neurons should decrease for those that are active in the opposing phase. The weight shouldn't alter if there is no signal correlation. A significant positive weight occurs between the nodes when their inputs are both positive and negative.

A significant negative weight exists between two nodes if one's input is positive while the input of the other is negative. All weights' values are zero at the beginning. It is possible to apply this learning rule for both soft- and hard-activation purposes. This is the unsupervised learning rule since the learning process does not use neurons' desired responses. The fact that the weights' absolute amounts are frequently proportionate to learning time is not acceptable.

**Mathematical Formulation**

According to the Hebbian learning rule, in every step of the algorithm the connection weight is improved by applying this rule:

$$\Delta w_{ji}(t) = \alpha x_i(t).y_j(t)$$

Here, $\Delta w_{ji}(t)$ = during time step $t$, the weight of the link rises

$\alpha$ = the positive and constant learning rate

$x_i(t)$ = the input value coming from the presynaptic neuron at time step $t$

$y_j(t)$ = the output of presynaptic neurons at the same time step $t$

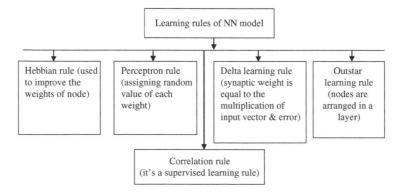

**FIGURE 3.17**
Types of learning rule of a NN model.

### 3.1.4.2 Perceptron Learning Rule

As you may already be aware, a neural network's connections each have a weight that varies as the network learns. It asserts that to demonstrate supervised learning, the network starts its learning process by assigning random values to each weight [28–30]. Calculate the output value based on a group of records for which we are aware of the expected output value. Following that, the network compares the estimated output value to the anticipated value.

**Mathematical Formulation**

Assume that there are $n$ finite input vectors, $x(n)$, and a desired or goal output vector, $T(n)$, where $n = 1$ to $M$, to understand its mathematical formulation. Now, as previously described, the output $Y$ may be determined based on the net input, and the activation function that is applied over that net input can be written as follows:

$$y = f(y_{in}) \quad = 1, y_{in} > \theta$$
$$= 0, y_{in} < \theta$$

Where $\theta$ is a threshold value and updating of the weight follows two different cases.

**Case 1.** When $T \neq Y$, then $w_{new} = w_{old} + t^*x$

**Case 2.** When $T = Y$, no change in weight

### 3.1.4.3 Delta Learning Rule

The delta rule is one of the most popular learning principles, invented by Widrow and Hoff. It depends on supervised learning. According to this rule, the multiplication of the error and the input results in the alteration of a node's sympatric weight [31]. It is a supervised learning algorithm in the sense that the activation function is continuous. This rule is based on the neverending gradient-descent method. The synaptic weights are altered by the delta rule in terms of reducing the net input to the output unit and the goal value.

**Mathematical Formulation**

The delta rule offered the updated the synaptic weights is given by

$$\Delta w_i = \alpha . x_i . e_j$$

Here,

$\Delta w_i$ = weight change for $i$th pattern

$\alpha$ = the positive and constant learning rate

$x_i$ = the input value from pre-synaptic neuron

$e_j = (t - y_{in})$ = the difference between the desired/target output and the actual output $y_{in}$

This delta rule is for a single output unit only; updating of the weight can be done by the following ways.

**Case 1.** When $t \neq y$, then $w_{new} = w_{old} + \Delta w$.

**Case 2.** When $t = y$, then no change in weight.

### 3.1.4.4 Competitive Learning Rule (Winner-takes-all)

There are certain unsupervised neural networks where the output nodes compete with one another to reflect the input pattern. We need to comprehend the competitive network in order to comprehend this learning rule [32]. A single layer feed forward network connected with feedback connections between outputs makes up this network. The competitors never sustain themselves because of the inhibitory type of linkages between outputs. The output nodes will compete with one another. The key idea is that the output unit that exhibits the maximum level of activation in response to a certain input pattern will be sorted the vector throughout training, due to the fact that just the winning neuron is altered whereas the losing neurons are kept intact; this rule is also known as the winner-takes-all rule.

**Mathematical formulation**

The three crucial elements for the mathematical formulation of this learning rule are as follows:

**Condition to be a winner**: A neuron $y_k$ considered as winner should satisfy following condition:

$$y_k = 1, \text{if } v_k > v_j \quad \begin{array}{l} \text{for all values of } j, j \neq k \\ = 0, \text{otherwise} \end{array}$$

This indicates $y_k$ wins the competition when its induced local field, say $v_k$, is the biggest among the network's other neurons.

**Condition of sum total of weight**: Another condition of this learning rule is algebraic. The cumulative weights assigned to a certain output neuron should be 1.

$$\sum_j w_{kj} = 1 \text{ for all } k$$

**Change of weight for winner**: When a neuron is dormant corresponding to the input pattern, it means learning rules didn't execute but for a wining neuron the corresponding output node is updated by the learning rule.

$$\Delta w_{kj} = -\alpha \left( x_j - w_{kj} \right) \quad \text{if neuron } k \text{ wins}$$
$$= 0, \quad \text{if neurons } k \text{ losses}$$

Where $\alpha$ is learning rate.

### 3.1.4.5 Outstar Learning Rule

As a result of known output in supervised learning, the Outstar learning rule was introduced by Grossberg. This rule is applied over the neurons grouped in a layer [33]. The layer of $p$ neurons is intended specifically to provide the required output.

**Mathematical formulation**: The final weight achieved by the rule is given by

$$\Delta w_j = \alpha \left( \mathrm{d} - w_j \right)$$

Here $d$ is the desired neuron output and $\alpha$ is the learning rate.

### 3.1.5 Mcculloch Pitts Neuron

This is the first mathematical model of a biological neuron invented by Warren Mcculloch and Walter Pitts in 1943. It has a linear threshold gate model and basic building blocks of a neural network [34, 35]. A directed weight graph used for connecting the neurons is shown in Figure 3.18. It has two possible status of neurons: active (logic 1) and silent (for logic 0).

### Architecture

Overall operation of this network segments into two parts. In the first part $I$ takes inputs, which multiply with individual weights and finally produce the output $y$. In second stage output indicates the two level decisions. For an example, to watch cricket or not on a TV is a binary input {0,1}, and the response is also binary, that is, 1 for will watch cricket and 0 for won't watch it.

### Concept/Condition:

Let threshold be denoted by $T$; then

$$Y = 1; \quad \text{when } X > T$$
$$= 0; \quad \text{when } X < T$$

Where $X = w_1 I_1 + w_2 I_2 + w_3 I_3 + \ldots + w_n I_n$

### Bias/Threshold:

This is the minimum value of weighted active input for a neuron to fire. If effective input (X) is larger than threshold value T, then output (Y) = 1, otherwise output (Y) =0.

Output $= f(a)$ [ where $a = \sum\limits_{i=1} w_i I_i - T$ and function $f$ where $\theta(I) = \begin{array}{l} 1, \text{ if } x > 0 \\ = 0 \text{ if } x < 0 \end{array}$ ]

### 3.1.6 Simple Neural Nets for Pattern Classification

The simplest architecture of neural network performs the pattern classification consists of a layer of inputs and a single output unit [36, 37]. Most of the neural net uses single layer architecture for pattern recognition is shown in Figure 3.19 & simplified neural network in Figure 3.20.

$$\text{net} = b + \sum_{i=1}^{n} x_i w_i \tag{3.1}$$

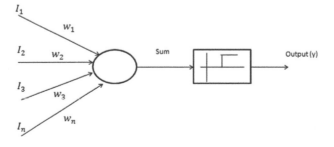

**FIGURE 3.18**
Architecture of Mcculloch Pitts neuron.

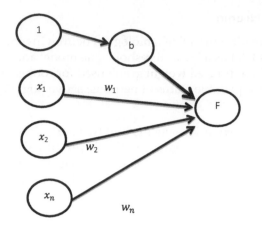

**FIGURE 3.19**
Single layer neural networks for pattern classification.

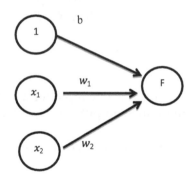

**FIGURE 3.20**
Simple neural networks.

**Biases and Threshold:**

A bias functions similarly to a weight on a connection from a unit whose activation is constant at 1. Improvements in the bias value correspond to rises in the net input value. In the event that a bias value is provided, the activation function is represented as

$$f(\text{net}) \quad = 1 \text{ if net} \geq 0$$
$$= -1 \text{ if net} \leq 0$$

If someone does not use the bias weight, then activation function can be expressed as

$$f(\text{net}) \quad = 1 \text{ if net} \geq \theta$$
$$= -1 \text{ if net} \leq \theta$$

$$\text{Net } \theta = \sum_{i=1}^{n} x_i w_i$$

Role of bias threshold: Here, we distinguish between areas of the input space with a positive net reaction and those with a negative net output.

$$\text{Net} = b + w_1 x_1 + w_2 x_2$$

The separation line separates the values of $x_1$ and $x_2$ for which the net responds positively from the values for which it responds negatively.

$$b + w_1 x_1 + w_2 x_2 = 0$$

$$x_2 = -\frac{w_1}{w_2} x_1 - \frac{b}{w_2}$$

To ensure that the net responds correctly to the training data, the values of $w_1$, $w_2$, and $b$ are established throughout the training process.

### 3.1.7 Linear Reparability

Two classes of patterns are considered to be linearly separable when they can be divided by a decision boundary and are represented by a linear equation.

In the previous figure in the $(x_1, x_2)$ plane these two classes $(x_1, x_2)$ can be separated by a single line L. They are known as a linearly separable pattern (Figure 3.21).

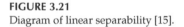

**FIGURE 3.21**
Diagram of linear separability [15].

### Linear separability for AND problem (Figure 3.22)

| A | B | Y |
|---|---|---|
| −1 | −1 | −1 |
| −1 | 1 | −1 |
| 1 | −1 | −1 |
| 1 | 1 | 1 |

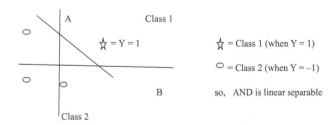

**FIGURE 3.22**
Linear separability for AND operation [15].

## Linear separability for OR problem (Figure 3.23)

| A | B | Y |
|---|---|---|
| −1 | −1 | −1 |
| −1 | 1 | 1 |
| 1 | −1 | 1 |
| 1 | 1 | 1 |

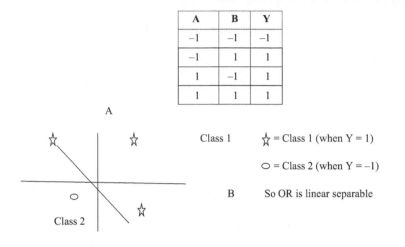

Class 1         ☆ = Class 1 (when Y = 1)

○ = Class 2 (when Y = −1)

So OR is linear separable

**FIGURE 3.23**
Linear separability for OR operation [16].

## Linear separability for XOR problem (Figure 3.24)

| A | B | Y |
|---|---|---|
| −1 | −1 | −1 |
| −1 | 1 | 1 |
| 1 | −1 | 1 |
| 1 | 1 | −1 |

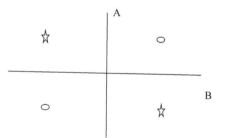

**FIGURE 3.24**
Linear separability for XOR gate [16].

So XOR logic is linear separable

### 3.1.8 Perceptron

This was developed by Rosenbalt utilizing the Mcculloch and Pitts model concept. The fundamental building block of an artificial neural network is the perceptron. It divides the labeled data into two classes.

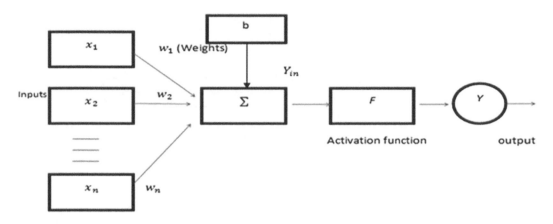

**FIGURE 3.25**
Typical architecture of perceptron.

## Characteristics of operation
It is made up of a single neuron with an unrestricted number of inputs and weight adjustments; however, the neuron's output can be either 1 or 0, depending on the threshold. This neuron's bias weight value is 1.

## Architecture
The simplified diagram is shown in Figure 3.25, and a description of each unit is explained in the next subsection.

## Basic elements of perceptron

Links: It would contain a number of interconnections, each of which has a weight, with the bias generally carrying a weight of 1.

Adder: After the input has been multiplied by the appropriate weights, the adder adds the input.

Activation function: This limits the neuron's output in the activation function. The Heaviside step function, which has two possible outputs, is the most fundamental activation function. If the input is positive, this method returns 1, and if it's negative, it returns 0.

## 3.2 Adaptive Linear Neuron (Adaline)

A network with only one linear unit is called Adaline, which stands for adaptable linear neuron. A bipolar activation function is used [38, 39]. To reduce the mean squared error (MSE) between the actual output and the desired output/target output, it applies the delta rule during training. Adjustable weights and bias are available. Adaline's fundamental design resembles a perceptron and includes an additional feedback loop that allows output to be compared to a desired or goal output [40]. The weights and bias will be modified following the comparison using the training process. The simplified diagram of Adaline is shown in Figure 3.26.

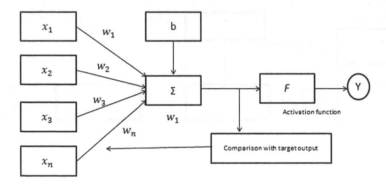

**FIGURE 3.26**
Basic diagram of adaline.

### 3.2.1 Multiple Adaptive Linear Neurons (Madaline)

A network made up of several Adalines parallel to one another is called Madaline, which stands for multiple adaptive linear neurons. It will only have one output [41, 42]. Adaline will function as a hidden unit between the input and Madaline layer, similar to a multilayer preceptor. The input and Adaline layers' weight and bias, as seen in the Adaline architecture, are both programmable. The constant bias and weights of the Madaline and Adaline layers are each 1. Delta rule can be used to aid with training. One neuron from the Madaline layer and $n$ neurons from the Adaline layer make up the Madaline architecture. Since it is situated between the output layer and the input layer, or the Madaline layer, the Adaline layer may be thought of as being a part of the hidden layer.

**Architecture**:

The $n$ input layer neurons, $m$ Adaline layer neurons, and one Madaline layer neuron make up the Madaline architecture. The Adaline layer, also known as the concealed layer, is positioned between the input layer and the output layer, or the Madaline layer. The simplified diagram of Madaline is shown in Figure 3.27.

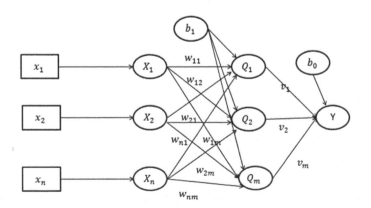

**FIGURE 3.27**
Basic architecture of Madaline.

### 3.2.2 Associative Memory Network

Based on the idea of pattern linkage, these neural networks are able to store a wide variety of patterns, and when producing an output, they choose one of the stored patterns by comparing it to the input pattern [43, 44]. The term "content-addressable memory" (CAM) is also used to describe these memories. With the stored patterns acting as data files, associative memory conducts a parallel search.

The two categories of associative memories that we can see are as follows:

- Auto associative memory
- Hetero associative memory

### 3.2.3 Auto Associative Memory

This neural network only has one layer, and both the input training vector and the output target vector are identical. In order for the network to hold a collection of patterns, the weights are chosen.

**Architecture**
The architecture of an auto associative memory network contains an equivalent number of output target vectors as input training vectors, as depicted in Figure 3.28.

### 3.2.4 Hetero Associative Memory

Similar to the auto associative memory network, this neural network has only one layer. The output target vector and the input training vector, however, are different in this network. In order for the network to hold a collection of patterns, the weights are chosen. There wouldn't be any nonlinear or delay operations feasible since hetero associative networks are static by nature.

#### 3.2.4.1 Architecture

There are $n$ input training vectors in a hetero associative memory network, and $m$ output target vectors, as depicted in the following image in Figure 3.29.

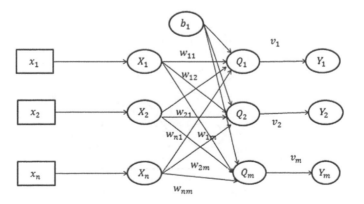

**FIGURE 3.28**
General diagram of auto associative memory.

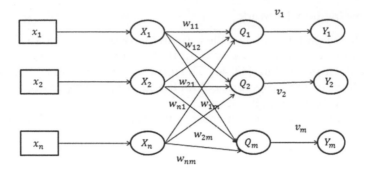

**FIGURE 3.29**
Basic diagram of hetero associative memory.

## 3.3 Bidirectional Associative Memory

Bidirectional associative memory (BAM) is an artificial neural network supervised learning paradigm. With hetero-associative memory, it can theoretically return a pattern of a different size given an input pattern. The human brain and this phenomenon are quite comparable. Association is a necessary part of human memory. It uses a series of mental associations, such as those between faces and names or between exam questions and answers, to help recover lost memories [45, 46]. A recurrent neural network (RNN) is required in such memory associations for one type of item with another in order to take an input pattern from one group of neurons and produce a similar but different output pattern from another set of neurons. Introducing such a network model has as its major goal the storage of hetero-associative pattern pairings. This is used to recover a pattern from an imperfect or noisy pattern.

**BAM architecture**:
BAM recalls $m$-dimensional vector Y from set B when given an input of an $n$-dimensional vector X from set A. The BAM also recalls X when Y is handled as input. The simplified bidirectional associative memory network is shown in Figure 3.30.

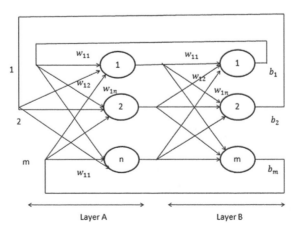

**FIGURE 3.30**
Basic diagram of bidirectional associative memory.

**Limitations of BAM:**

Storage capacity of the BAM: The number of associations that can be kept in the BAM should not be greater than the number of neurons in the lower layer.

Erroneous convergence: BAM might not always produce the relationship that is closest.

---

## 3.4 Self-Organizing Maps: Kohonen Maps

A self-organizing map is an additional kind of artificial neural network (also known as a Kohonen map or SOM) was further inspired by 1970s-era biological models of neuronal networks. Its network was trained using a competitive learning algorithm and an unsupervised learning approach. It is a specific sort of ANN that draws inspiration from biological representations of brain systems. In order to simplify the problem and make interpretation simple, multidimensional data are mapped to lower dimensional data. The structure of the SOM with $n$ input characteristics and two clusters for every sample is given in Figure 3.31.

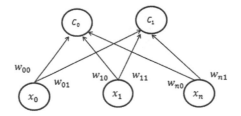

**FIGURE 3.31**
Simple diagram of self-organization map.

## ALGORITHM

**Step 1:** Initialize the weights $w_{ij}$ random values may be assumed. Initialize the learning rate $\alpha$.

**Step 2:** Calculate the square of Euclidian distance, i.e,. for each $j = 1$ to $m$

$$D(j) = \sum_{i=1}^{n}\sum_{j=1}^{m}(X_i - w_{ij})^2$$

Where $i =$ is number of inputs, $j =$ number of cluster in input section, and $w_{ij}$ is a weight of each cluster

**Step 3:** Find winning unit index $j$, so that D (j) is minimum

**Step 4:** For all units $j$ within a specific neighborhood $j$ and for all $i$ calculate new weight.

$$(w_{ij})\text{new} = (w_{ij})\text{old} + \alpha\left[X_i - (w_{ij})\text{old}\right]$$

**Step 5:** update learning rate $\alpha$ using formula

$$\alpha(t+1) = 0.5\,\alpha(t)$$

## 3.5 Learning Vector Quantization (LVQ)

This is a particular kind of ANN that was also motivated by biological models of the neural network. It is built on a classification method for supervised learning and developed its network using a competed reinforcement learning like a self-organizing map. It can resolve the multiclass issue [47, 48]. There are two layers in LVQ: an input layer and an output layer, as shown in Figure 3.32.

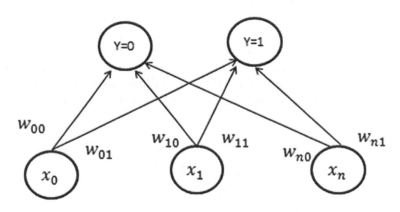

**FIGURE 3.32**
Simple diagram of learning vector quantization.

### ALGORITHMS

**Step 1:** Create a reference vector by starting with a collection of training vectors, using the first "$n$" (the number of clusters) training vectors as weight vectors while still preserving the training potential of the leftover training vectors.

**Step 2:** Calculate the Euclidean distance for $i = 1$ to $n$ and $j = 1$ to $m$

$$D(j) = \sum_{i=1}^{n}\sum_{j=1}^{m}\left(X_i - w_{ij}\right)^2$$

Finding winning unit index $D(j)$ which has minimum value.

**Step 3:** Update weights or winning unit "$w_i$" using following condition

$$\text{If } T = J; \left(w_{ij}\right)\text{new} = \left(w_{ij}\right)\text{old} + \alpha\,[X_i - \left(w_i\right)\text{old}]$$
$$\text{If } T \neq J; \left(w_{ij}\right)\text{new} = \left(w_{ij}\right)\text{old} - \alpha\,[X_i - \left(w_i\right)\text{old}]$$

Where, $T$ = target vector and $J$ = winning set vector

## 3.6 Counter Propagation Network (CPN)

In a counter propagation network there are several combinations that take place between the input, output, and cluster layer. A counter propagation network is constructed with the help of an instar and outstar model [49]. The instar–outstar model is a three layer model where input and output mapping was done by using the learning algorithm to produce output vector Y corresponding to input vector X. The CPN model has two stages; in the first stage input vectors are clustered, while in the second stage, to obtain the output, determine the weight of the cluster layer [50, 51]. In CPN there are three layers, namely input layer, output layer, hidden layer. The input layer is also known as the Kohonen layer the output layer is known as the Grossberg layer.

### Classification

### 3.6.1 Full Counter Propagation Network (FCPN)

This CPN creates a look up table. In the look up table there several combinations of X: Y vectors are available. It works effectively with the inverse function. The simplified diagram of FCPN is shown in Figure 3.33.

### 3.6.2 Forward Only Counter Propagation Network

This is a simplified form of FCPN as shown in Figure 3.34. In this CPN, it only makes a cluster at Kohonen units. Initially weights are trained which are connected between the input layer and cluster layer, and then weights are trained between the output layer and cluster layer. Here the target of the network should be knowledge.

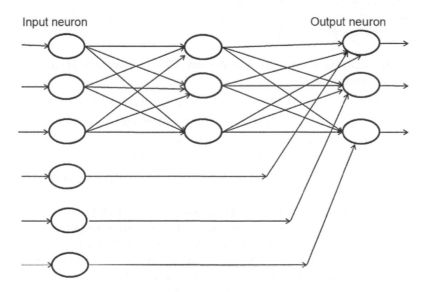

**FIGURE 3.33**
Basic architecture of full counter propagation network.

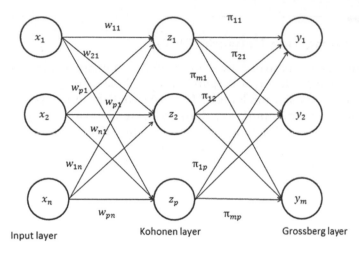

**FIGURE 3.34**
Basic architecture of forward only counter propagation network.

## 3.7 Adaptive Resonance Theory (ART)

In 1987, Stephen Grossberg and Gail Carpenter introduced this network. It has two new features: adaptiveness (open to learning) and resonance (preserving prior knowledge). The unsupervised learning approaches are used in the fundamental ART. An approach for clustering is applied by ART networks. The network and algorithm are shown the input.

### Types of Adaptive Resonance Theory (ART)
After 20 years of study, Carpenter and Grossberg created many ART architectures. The following categories apply to ARTs.

- ART1 is the most basic and straightforward ART architecture. It can cluster input values with binary data.
- ART2 is an extension of ART1 that can cluster input data with continuous values.
- Fuzzy ART is the combination of fuzzy logic and artificial intelligence.
- ARTMAP—this supervised method of ART learning allows one ART to build on knowledge from a prior ART module. Predictive ART is another name for it.
- FARTMAP—this supervised ART architecture also incorporates fuzzy logic.

### Fundamental Architecture of Adaptive Resonance Theory (ART)
A competitive, self-organizing neural network is the adaptive resonant theory. It can be either the supervised or unsupervised types (ART1, ART2, ART3, etc.), for ARTMAP the name of the supervised algorithms typically ends in "MAP". However, the fundamental ART model is unsupervised in nature and is made up of the F1 layer, also known as the comparison field, and the F2 layer, also known as the recognition field (which consists of the clustering units) resetting module (that acts as a control mechanism) as shown in Figure 3.35.

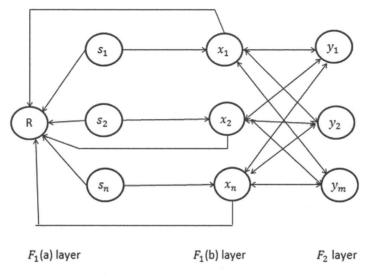

**FIGURE 3.35**
Basic architecture of adaptive resonance theory (ART).

The input layer $F_1(a)$ of the comparison layer is made up of units $s_1, s_2...s_n$ while the interface layer $F_1(b)$ is made up of units $x_1, x_2,....x_n$. The input pattern simply passes it on to interface layer $F_1(b)$ but does not proceed to the input layer $F_1(a)$. That is $[s_1, s_2...s_n]$ connects to $[x_1, x_2,... x_n]$. The interface layer $F_1(b)$ is responsible for transmitting the input pattern to the recognition layer $F_2$, which compares the input pattern with the winning code vector. The $F_2$ Recognition layer is also known as competitive layer.

A unique cluster is represented by each unit of the $F_2$ layer. The number of units in the $F_2$ layer is not fixed; however, ART1 offers the possibility of increasing the number of clusters. The $F_2$ unit may be in any of the following three states as the ART learns a pattern: (1) Active: This unit has a positive activation and is "on." (2) Inactive: Activation = 0 and the unit is "off." But this team participates in competition. (3) This device's activation is zero, making it "off." Additionally, when learning with the current input pattern, it is not permitted to continue competing.

## Advantages of ART

- Stable and not distributed with different other ways to get good results
- Integrated and used with a large variety of inputs offered to its network
- Competition learning is incapable of adding new clusters when essential

## Applications

- Mobile robot control
- Face recognition
- Land recover classification
- Medical diagnosis
- Signature verification

## 3.8 Standard Back-Propagation Architecture

Application of this learning technique to a multilayer feed forward network with a continuous differentiable function is shown in Figure 3.36. It also uses gradient descent with a differentiable function. In this method error is propagated back to the hidden layer [52–54]. The training of the BPN network is done in three stages: feed forward, back propagation of error, and weight updater.

### Architecture

Neurons present on the hidden and output layer have biases equal to 1. During feed forward information flows in the forward direction. During back propagation output signals are sent back. Any function that rises monotonically and is differentiable might serve as an activation function. Error generated at the terminal end is back propagated to the output layer and hidden layer to adjust the weight value.

To solve a linear separable problem, use more than one perceptron. Combining their output into another perceptron would produce a final indication. For a perceptron in the first layer the input comes from the actual inputs while the perceptron present in the second layer gets input from output of the first perceptron. The perceptron of the second layer cannot distinguish whether the actual inputs from first layer were on or off. A hard-hitting threshold function removes information that is needed for further training.

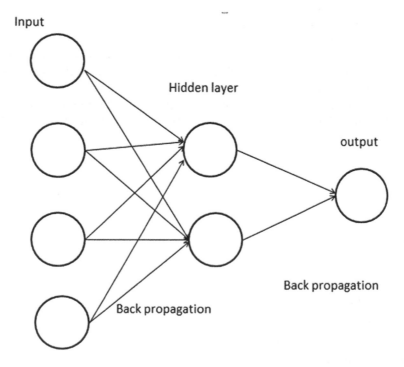

**FIGURE 3.36**
Basic architecture of back-propagation neural network.

## 3.9 Boltzmann Machine Learning

Boltzmann machine learning is a stochastic learning processes, which is the foundation of the prior optimization methods utilized in ANN and has a recurrent nature. Geoffrey Hinton and Terry Sejnowski created the Boltzmann machine in 1985. The clarity of this algorithm provided by Hinton on the Boltzmann machine is more precise [55]. "This network's utilization of only locally available information is a surprise characteristic. Although the modification improves a global metric, the weight change only has an impact on how the two units it links behave, according to Ackley and Hinton in 1985. Some of the important characteristics of the recurrent structure of Boltzmann machined are as follows [56].

- They are made up of stochastic neurons, each of which can exist in either one of two conceivable states, 1 or 0.
- This neuronal network has both clamped and adaptable free state neurons.
- Simulated annealing would transform a discrete Hopfield network into a Boltzmann machine.

**Objective of Boltzmann Machine**
Maximizing a problem's remedy is the fundamental objective of a Boltzmann system. Optimizing the weights and quantities in respect to that particular problem is the role of the Boltzmann machine.

**Architecture**
The Boltzmann machine's construction is depicted in Figure 3.37. The diagram makes it apparent that the array of units is two-dimensional. Weights on connections between units in this case are $p$ when $p > 0$, and weights of self-connections are provided by $b$, where $b > 0$.

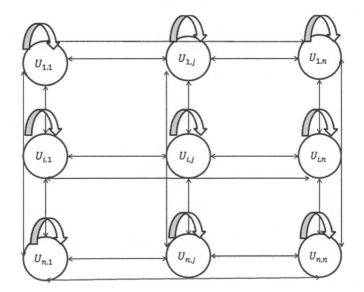

**FIGURE 3.37**
Basic architecture of Boltzmann machine learning.

## References

1. Abiodun OI, Jantan A, Omolara AE, Dada KV, Mohamed NA, Arshad H. State-of-the-art in artificial neural network applications: A survey. *Heliyon* 2018;4(11): 1–41.
2. Wang SC. Artificial neural network. In *Interdisciplinary computing in java programming*, Springer;2003, pp. 81–100.
3. Vohradsky J. Neural network model of gene expression. *The FASEB Journal* 2001;15(3): 846–854.
4. Hu YH, Hwang JN (Eds.). *Handbook of neural network signal processing*. Springer;2002.
5. Bebis G, Georgiopoulos M. Feed-forward neural networks. *IEEE Potentials* 1994;13(4): 27–31.
6. Miyamoto H, Kawato M, Setoyama T, Suzuki R. Feedback-error-learning neural network for trajectory control of a robotic manipulator. *Neural Networks* 1988;1(3): 251–265.
7. Kawato M. Feedback-error-learning neural network for supervised motor learning. In *Advanced neural computers*, Eckmiller Rolf (Eds.). North-Holland;1990, pp. 365–372.
8. Abiodun OI, Jantan A, Omolara AE, Dada KV, Mohamed NA, Arshad H. State-of-the-art in artificial neural network applications: A survey. *Heliyon* 2018;4(11): e00938.
9. Caruana R, Niculescu-Mizil A. An empirical comparison of supervised learning algorithms. In *Proceedings of the 23rd International Conference on Machine Learning*. 2006, June, pp. 161–168.
10. Saravanan R, Sujatha P. A state of art techniques on machine learning algorithms: a perspective of supervised learning approaches in data classification. In *2018 Second International Conference on Intelligent Computing and Control Systems (ICICCS)*. IEEE;2018, June, pp. 945–949.
11. Oliver A, Odena A, Raffel CA, Cubuk ED, Goodfellow I. Realistic evaluation of deep semi-supervised learning algorithms. *Advances in Neural Information Processing Systems* 2018;31. https://doi.org/10.48550/arXiv.1804.09170
12. Celebi ME, Aydin K (Eds.). *Unsupervised learning algorithms*. Springer International Publishing;2016.
13. Palacio-Niño JO, Berzal F. Evaluation metrics for unsupervised learning algorithms. 2019. arXiv:1905.05667.
14. Ghahramani Z. Unsupervised learning. In *Summer school on machine learning*. Springer;2003, February, pp. 72–112.
15. Hastie T, Tibshirani R, Friedman J. Unsupervised learning. In *The elements of statistical learning*. Springer;2009, pp. 485–585.
16. Oh J, Hessel M, Czarnecki WM, Xu Z, van Hasselt HP, Singh S, Silver D. Discovering reinforcement learning algorithms. *Advances in Neural Information Processing Systems* 2020;33: 1060–1070.
17. Szepesvári C. Algorithms for reinforcement learning. *Synthesis Lectures on Artificial Intelligence and Machine Learning* 2010;4(1): 1–103.
18. Jordan S, Chandak Y, Cohen D, Zhang M, Thomas P. Evaluating the performance of reinforcement learning algorithms. In *International Conference on Machine Learning*. PMLR;2020, November, pp. 4962–4973.
19. Singh S, Jaakkola T, Littman ML, Szepesvári C. Convergence results for single-step on-policy reinforcement-learning algorithms. *Machine Learning* 2000;38(3): 287–308.
20. Sharma S, Sharma S, Athaiya A. Activation functions in neural networks. *Towards Data Science* 2017;6(12): 310–316.
21. Agostinelli F, Hoffman M, Sadowski P, Baldi P. Learning activation functions to improve deep neural networks. 2014. arXiv:1412.6830.
22. Zhang H, Weng TW, Chen PY, Hsieh CJ, Daniel L. Efficient neural network robustness certification with general activation functions. *Advances in Neural Information Processing Systems* 2018;31: 4944–4953.
23. Kheradpisheh SR, Masquelier T. Temporal backpropagation for spiking neural networks with one spike per neuron. *International Journal of Neural Systems* 2020;30(06): 2050027.

24. Hazan H, Saunders DJ, Khan H, Patel D, Sanghavi DT, Siegelmann HT, Kozma R. Bindsnet: A machine learning-oriented spiking neural networks library in python. *Frontiers in Neuroinformatics* 2018;12: 89.

25. Zhang Y, Tiňo P, Leonardis A, Tang K. A survey on neural network interpretability. *IEEE Transactions on Emerging Topics in Computational Intelligence* 2021;5: 726–742.

26. Chakraverty S, Sahoo DM, Mahato NR. Hebbian learning rule. In *Concepts of soft computing*. Springer;2019, pp. 175–182.

27. Amato G, Carrara F, Falchi F, Gennaro C, Lagani G. Hebbian learning meets deep convolutional neural networks. In *International Conference on Image Analysis and Processing*. Springer;2019, September, pp. 324–334.

28. Moldwin T, Segev I. Perceptron learning and classification in a modeled cortical pyramidal cell. *Frontiers in Computational Neuroscience* 2020;14: 33.

29. Gobinath S, Madheswaran M. Deep perceptron neural network with fuzzy PID controller for speed control and stability analysis of BLDC motor. *Soft Computing* 2020;24(13): 10161–10180.

30. Zhang M, Wu J, Belatreche A, Pan Z, Xie X, Chua Y, … Li H. Supervised learning in spiking neural networks with synaptic delay-weight plasticity. *Neurocomputing* 2020;409: 103–118.

31. Frazier-Logue N, Hanson SJ. Dropout is a special case of the stochastic delta rule: Faster and more accurate deep learning. 2018. arXiv:1808.03578.

32. Qu L, Zhao Z, Wang L, Wang Y. Efficient and hardware-friendly methods to implement competitive learning for spiking neural networks. *Neural Computing and Applications* 2020;32(17): 13479–13490.

33. Wilamowski BM. Neural networks learning. In *Intelligent systems*. CRC Press;2018, pp. 11–17.

34. Chen Y, Tang Z, Todo H. New mechanism of visual motion direction detection based on Mcculloch-Pitts neuron model. In *2021 4th International Conference on Artificial Intelligence and Big Data (ICAIBD)*. IEEE;2021, May, pp. 448–453.

35. Chakraverty S, Sahoo DM, Mahato NR. McCulloch–Pitts neural network model. In *Concepts of soft computing*. Springer; 2019, pp. 167–173.

36. Wang Z, Joshi S, Save'lev S, Song W, Midya R, Li Y, … Yang JJ. Fully memristive neural networks for pattern classification with unsupervised learning. *Nature Electronics* 2018;1(2): 137–145.

37. Nakazawa T, Kulkarni DV. Wafer map defect pattern classification and image retrieval using convolutional neural network. *IEEE Transactions on Semiconductor Manufacturing* 201831(2): 309–314.

38. Lau MM, Lim KH. Review of adaptive activation function in deep neural network. In *2018 IEEE-EMBS Conference on Biomedical Engineering and Sciences (IECBES)*. IEEE;2018, December, pp. 686–690.

39. Zamora J, Rhodes AD, Nachman L. Fractional adaptive linear units. In *Proceedings of the AAAI Conference on Artificial Intelligence*. 2022, June, Vol. 36, No. 8, pp. 8988–8996.

40. Costa AC, Ahamed T, Stephens GJ. Adaptive, locally linear models of complex dynamics. *Proceedings of the National Academy of Sciences* 2019;116(5): 1501–1510.

41. Al-Mohdar AA, Bahashwan AA. Investigating sensitivity of nonlinear classifiers by reducing mean square error (MSE). *International Research Journal of Innovations in Engineering and Technology* 2020;4(8): 31.

42. Handayani AN, Aindra AD, Wahyulis DF, Pathmantara S, Asmara RA. Application of adaline artificial neural network for classroom determination in elementary school. In *IOP Conference Series: Materials Science and Engineering*. IOP Publishing;2018, November, Vol. 434, No. 1, p. 012030.

43. Shriwas R, Joshi P, Ladwani VM, Ramasubramanian V. Multi-modal associative storage and retrieval using Hopfield auto-associative memory network. In *International Conference on Artificial Neural Networks*. Springer, Cham; 2019, September, pp. 57–75.

44. Sun J, Han G, Zeng Z, Wang Y. Memristor-based neural network circuit of full-function pavlov associative memory with time delay and variable learning rate. *IEEE Transactions on Cybernetics* 2019;50(7): 2935–2945.

45. Humphries U, Rajchakit G, Kaewmesri P, Chanthorn P, Sriraman R, Samidurai R, Lim CP. Global stability analysis of fractional-order quaternion-valued bidirectional associative memory neural networks. *Mathematics* 2020;8(5): 801.

46. Li Y, Li J, Li J, Duan S, Wang L, Guo M. A reconfigurable bidirectional associative memory network with memristor bridge. *Neurocomputing* 2021;454: 382–391.

47. Pourghasemi HR, Gayen A, Lasaponara R, Tiefenbacher JP. Application of learning vector quantization and different machine learning techniques to assessing forest fire influence factors and spatial modelling. *Environmental Research* 2020;184: 109321.

48. Sarhan S, Nasr AA, Shams MY. Multipose face recognition-based combined adaptive deep learning vector quantization. *Computational Intelligence and Neuroscience* 2020;2020: 8821868.

49. Singh UP, Jain S, Tiwari A, Singh RK. Gradient evolution-based counter propagation network for approximation of noncanonical system. *Soft Computing* 2019;23(13): 4955–4967.

50. Agrawal S, Singh RK, Singh UP, Jain S. Biogeography particle swarm optimization based counter propagation network for sketch based face recognition. *Multimedia Tools and Applications* 2019;78(8): 9801–9825.

51. Kaden M, Schubert R, Bakhtiari MM, Schwarz L, Villmann T. The LVQ-based counter propagation network: An interpretable information bottleneck approach. In *ESANN – 2021 Proceedings*. 2021, 6–8 October.

52. da Silva LEB, Elnabarawy I, Wunsch II DC. A survey of adaptive resonance theory neural network models for engineering applications. *Neural Networks* 2019;120: 167–203.

53. da Silva LEB, Elnabarawy I, Wunsch II DC. Dual vigilance fuzzy adaptive resonance theory. *Neural Networks* 2019;109: 1–5.

54. da Silva LEB, Elnabarawy I, Wunsch II DC. Distributed dual vigilance fuzzy adaptive resonance theory learns online, retrieves arbitrarily-shaped clusters, and mitigates order dependence. *Neural Networks* 2020;121: 208–228.

55. Kiraly B, Knol EJ, van Weerdenburg WM, Kappen HJ, Khajetoorians AA. An atomic Boltzmann machine capable of self-adaption. *Nature Nanotechnology* 2021;16(4): 414–420.

56. Decelle A, Furtlehner C. Restricted Boltzmann machine: Recent advances and mean-field theory. *Chinese Physics B* 2021;30(4): 040202.

# 4

## Introduction to Genetic Algorithm

### 4.1 Introduction

Every process or model designed is centered on options aimed at achieving one or more objectives. The study of optimization focuses on how to mathematically model this process and, within the parameters of these models, how to make the best decisions [1–3]. Making anything better is the process of optimization. In any mathematical model of a process, there will be a set of inputs and a set of outputs as shown in Figure 4.1.

Finding the input values that provide the "best" output values is referred to as optimization. In different contexts, the word "best" can imply different things, but in mathematics it often refers to increasing or decreasing one or more objective functions by adjusting the input parameters [4]. The whole range of possible outcomes or values for the inputs makes up the search space. The best answer in this search region can be found at a certain point or set of points. Finding that point or those points in the search space is the goal of optimization. Every optimization problem consists of three components: an objective function, constraints, and choice variables, as shown in Figure 4.2.

The phrase "formulating an optimization problem" refers to the process of turning a real-world challenge into the three categories of the quantitative equations and parameters [5]. The objective function, which is frequently represented by the letters "$f$" or "$z$," reflects a single quantity that can be maximized or minimized. The terms "minimize expense," "maximize flow rate," "maximize output voltage," "minimize material removal rate," and others are examples used in different process industries. From the literature survey, it is seen that there is no chance of simultaneously optimizing several objectives without understanding how to maximize a single objective function. It is best to understand the fundamentals of optimization in a simpler setting before moving on to more complicated multi-objective optimization approaches [6, 7].

The vector $x$ represents the decision variables, which indicate the components of the situation that are under control. Both variables you can directly pick and variables you can indirectly impact through the selection of other decision variables can be included in this. For instance, a variety of independent factors, such as liquid properties, pipe diameter, type of flow sensor (contact type or non-contact type), and temperature of the process plant, must depend upon one another in order to achieve the best flow rate (objective function). Each independent variable in formulation should have the potential to either directly affect the objective function or indirectly affect the objective function through other decision variables [8–10].

DOI: 10.1201/9781003216001-5

**FIGURE 4.1**
Basic diagram of any process.

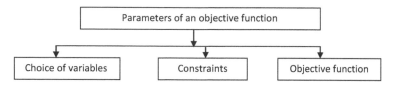

**FIGURE 4.2**
Components of optimization techniques.

Any type of restriction on the values that the decision variables take is referred to as a constraint. The most obvious and direct constraints are those that directly restrict your options, such as laws requiring you to maintain equipment to a certain standard, restrictions on changing the highest flow rate allowed in the process industry, and requirements that each independent and dependent variable have a minimum and maximum range. As an illustration, let's say you 60 feet of fence to work with and want to surround the greatest rectangular area; what size should the space be that is fenced off?

In the first stage, mathematical notation can be done. Let $L$ and $W$ represent the choice variables, length and width, respectively. When the area is maximized, which is equal to $LW$, the objective function is maximized. The perimeter limitation can be written as $2L + 2W \leq 60$. Finally, independent variables have nonnegative constraints, i.e., $L \geq 0$ and $W \geq 0$. All of this information is concisely represented in the following way:

Max: $LW$
Subject to $2L + 2W \leq 60$
$L \geq 0 \; W \geq 0$

## 4.2 Optimization Problems

There are many techniques applied to accomplish the objective of a mathematical model. The majority of the time, this denotes qualities like determining a function's maximum or lowest value, the shortest amount of time necessary to compute the task, the least amount of money spent on the goal function, the most power needed by the device to accomplish the work, and so forth [11–13]. These kinds of issues can be resolved by identifying the proper function, followed by methods for determining the maximum or minimum value of the predictor variables or attribute values. Typically, a problem of this kind will have the following mathematical form: When $a \leq x \leq b$ or sometimes $a$ or $b$ is infinite, find the biggest (or smallest) value of $f(x)$. The interesting part of this research is, in the domain of $a$ and $b$, what are smallest and largest values of the function $f(x)$.

### 4.2.1 Steps for Solving the Optimization Problem

1. Identification of maximized or minimized value of the objective function and the constraints for the present research.
2. Labeling the input variables and output variables or objective functions.
3. Labeling the units of each input and output variable, for example, $D$ for the diameter of a PVC pipe in centimeters or $\rho$ for the density of the liquid in grams per cubic centimeter.
4. Formulating the objective functions to be maximized or minimized.
5. Expressing objective function formulation by means of independent variables.
6. Applying different computational algorithms to achieve the optimum objective function by keep in mind the constraints of the independent variables. For an example, determine the sides of a 100 square unit rectangle having the shortest perimeter. The following methods have to be implemented to obtain the characteristics of the polygon. First, this condition turns into a problem that is entirely mathematical and requires us to determine the lowest value of the objective function of the perimeter of a rectangle.

Let $x$ denote one of the sides of the rectangle; then the value of the adjacent side is $\dfrac{100}{x}$.
The function that we aim to reduce is hence $f(x) = 2*x + 2*\dfrac{100}{x}$. For, $x > 0$ and $\dfrac{100}{x} .> 0$
Next we find the value of $f(x)$; $f'(x)$ should be zero.

$$f'(x) = 2 - \frac{200}{x^2} = 0$$

Solving, we get $x = \ddagger\, 10$ is of interest.

Since $f'(x)$ is specified throughout on the range $(0, \infty)$, and neither critical values nor end points remain. To identify a relative maximum, minimum, or neither maximum or minimum value, the second-order derivative is needed.

$f''(x) = \dfrac{400}{x^3} > 0$ for $x = 10$; hence, there is a relative minimum value.

As there is only a critical value, the smallest perimeter of the rectangle is $= 2 * (10 + 10) = 40$ units, and it is only possible when this polygon is a square.

### 4.2.2 Point-to-Point Algorithms (P2P)

Recently, there has been a lot of interest in this basic optimization problem, and significant progress has been made. After giving a brief history of earlier findings, this subsection highlights the contemporary heuristics algorithm that addresses the issue by looking at just a piece of the input graph (quite large mapping) to solve the optimization problem [14]. Point-to-point algorithms identify precise shortest paths. These algorithms are heuristic since they only work well with specific classes of graphs. Although they have performed well in experiments, there are no known theoretical constraints that can explain the results. The majority of these algorithms are driven by the desire to locate routes via extensive road networks. P2P algorithms are the combination of two logical algorithms, the traditional Dijkstra's algorithm and its bidirectional variant, which were created in the 1950s and 1960s, respectively.

In the P2P algorithm, the overall process is formulated in this way: Find the shortest route between the source vertex, $s$, and the destination vertex, $t$, on the directed graph $G = (V, A)$, which has nonnegative arc lengths [15, 16]. Exact shortest pathways are what we're after. Limit the amount of the precomputed data in the preprocessing step to a constant multiplied by the size of the input graph. Practical factors impose a time restriction on the preprocessing phase.

### 4.2.3 A∗ Search Algorithm

The $A^*$ search method was initially proposed to accelerate search in large implicitly represented game graphs. i* search algorithm is also known as goal-directed search or heuristic search [17–19]. The concept is to direct the search toward because one biases the search toward rather than seeking a ball around $s$. The algorithm uses a (perhaps domain-specific) function $\pi_t : V \to R$ such that $\pi_t(v)$ gives an estimate on the distance from $v$ to $t$. Define a (forward search) key of $v, k_s(v) = d_s(v) + \pi_t(v)$. Only one thing separates the A* search from Dijkstra's algorithm: at each step, the former chooses a labeled vertex $v$ with the smallest key to scan next as opposed to the one with the smallest $d(s)$ value. On a network with a length function of $\pi_t$, if $\pi_t$, then it is clear that an $A^*$ search is identical to Dijkstra's method. The algorithm is accurate if it is viable and nonnegative.

### 4.2.4 Simulated Annealing

Hill climbing allows only an upward direction, but simulated annealing allows the downward steps [20]. It follows the global maxima. It checks the value of the entire neighborhood. Simulated annealing simulates metallurgical techniques, which include heating a material to an elevated temperature and then cooling it. As the material cools to become a pure crystal, impurities are frequently eliminated as a result of atoms shifting unexpectedly at high temperatures. The simulated annealing optimization process replicates this, with the energy state correlating to the present solution [21, 22].

We provide a beginning temperature, which is typically set to 1, and a limiting temperature, on the order of $10^{-4}$. In order to reduce the temperature till it exceeds the required temperature, the current temperature is multiplied by some proportion alpha. We perform the core optimization process a predetermined number of times for each unique temperature value. Finding an adjacent solution and accepting it with a probability of $e^\wedge((f(c)–f(n)))$ is what the optimization method entails, where $c$ stands for the present solution and $n$ for the surrounding solution. The present solution is somewhat perturbed to find a neighboring solution. This unpredictability helps avoid the usual problem when optimization strategies become stuck in local minima. The method is more probable to evolve close to the global optimum by conceivably embracing a less ideal solution than the one we now have and admitting it with a probability inverse to the increase in cost. Figure 4.3 depicts how simulating annealing's goal function behaves in relation to the state space.

The design of a neighbor function can be challenging and must be done on a case-by-case basis, although the principles listed subsequently can help in identifying neighbors in location optimization situations. Move every point 0 or 1 units, randomly distributed; shuffle the supplied pieces at random. Alternately, for random input sequence elements, input sequence permutation. Divide the input sequence into random segments, then permute the segments.

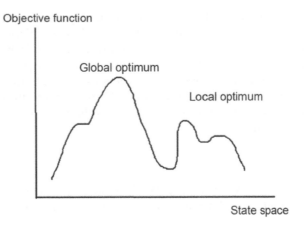

**FIGURE 4.3**
Objective function versus state space of simulated annealing.

**Advantages:**

- Easy to code for complex problems.
- Easy to give the best solution.
- Statistically guarantees finding globally optimum solution.

**Disadvantages:**

- This algorithm runs very slowly.
- This algorithm never conveys the information that it achieved the optimum solutions.

### 4.2.5 Genetic Algorithm (GA)

Algorithms for optimization that draw inspiration from biology are known as genetic algorithms. Evolution is a theory put out by Charles Darwin that explains how species grow biologically via mating preference and the survival of the fittest [23–25]. Deoxyribonucleic acid (DNA) is a representation created through evolution. Evolutionary processes are built on the DNA, which encodes organisms. The uninterrupted cycle of artificial development based on the ideas of natural evolution is shown in Figure 4.4. Starting with random or purposefully initiated solutions, the evolutionary process begins. The crossover operator is used to combine two or more solutions once again to begin the evolutionary cycle, which

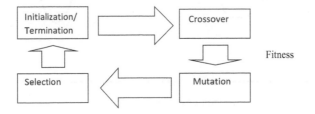

**FIGURE 4.4**
Genetic algorithm complete cycle.

ultimately results in a mutant result. For the subsequent generation, the finest results from this generation are chosen. The adaptive cycle then checks to see if the stopping criteria has been satisfied and, if not, repeats the genetic optimization run. However, the most basic form of genetic algorithms only uses one parent who is altered to produce a kid. The parent or child is chosen by the selection operator as the superior option. Recombination is not used because each generation has a single parent. Crossover and mutation operators may be created for practically all resolution formulations. Figure 4.4 represents the complete cycle of the genetic algorithm.

### 4.2.5.1 Motivation of GA

Genetic algorithms are appealing for use in addressing optimization issues because they can yield faster and better results than other algorithms. The following list outlines the requirements for Gas.

1. Solving Complex Problems
   During a large set of problems, numerical programming on even a powerful computing system takes a longer time to execute the problem; by finding the solution in such cases, GA proves its efficiency for proving the optimal output within a short period of time.
2. Failure of Gradient Based Methods
   For traditional algorithms like the hill-climbing method, simulated annealing, and others, the process starts by considering an initial point and moving toward the direction of a top of the hill. All these algorithms are suitable for the single objective cost function, but in most complex problems there are a number of peaks and many valleys, which makes this algorithm produce an unstable optimum value as shown in Figure 4.5.

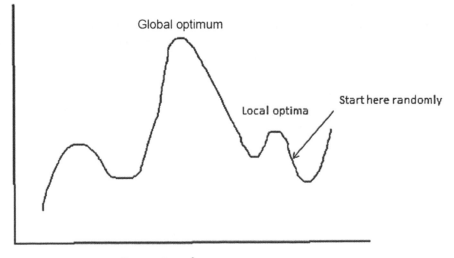

**FIGURE 4.5**
Objective function of a gradient-based searching method.

3. Getting a Good Solution Fast

There are some real complex problems like the travelling salesperson problem (TSP), optimum power flow, and very large scale integrated circuit VLSI design available in research fields as common examples. To solve such real problems using a computational algorithm, taking longer time is never acceptable; therefore a good enough fast solution is required.

### 4.2.5.2 Basic Terminology

It is important to be familiar with certain basic terms that will be used throughout applications of genetic algorithms. Figure 4.6 explain different metrics of the genetic algorithm.

**Population**—Possible encoded solution for a given problem. The nature of the answer is comparable to the number of people.

**Chromosomes**—This is a solution to the given problem.

**Gene**—Elementary position of a chromosome and its value defined by allele.

**Genotype and phenotype**—Genotype is the population in the computation space manipulated by a computing system, while phenotype indicates the populations of real-world problems.

**Decoding and encoding**—This is a vice versa process for conversion between genotype and phenotype. For a given problem, decoding is used to convert the genotype to phenotype, and to convert phenotype to genotype, encoding techniques are used. Both the techniques are used to calculate the fitness value effectively.

**Fitness function**—This is a function to produce the suitable solution for a given set of inputs.

**Genetic operators**—This is a genetic composition to solve an objective function. It includes crossover, mutation, selection, and so on. In next section, each of the genetic operators is explained.

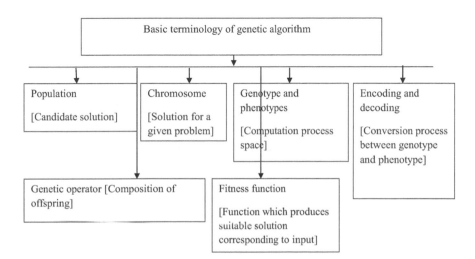

**FIGURE 4.6**
Basic metrics of genetic algorithm.

### 4.2.5.2.1 Crossover

Crossover allows producing two or more solutions after mating of two or more genetic materials. Generally most of the organism has parents. Some exceptions only have one parent since they are unaware of the existence of other sexes [26, 27]. The choice of a potential mate partner is the first stage in nature. Many species invest a lot of energy in the selection of a partner and try to apply different techniques to attract the partner in a process. The second stage, after proper selection of mate is paring. From a biological standpoint, two individuals of the same species mix their genetic properties and produce offspring. The technique used by crossover operators in genetic algorithms combines the genetic properties of the parents. N-point crossover is a popular one for bit string representation. It alternately separates two solutions at positions $n$ and reassembles them into a new one as shown in Figure 4.7.

Solution of the two operators is outperformed after their best properties are combined by the mating. The new generated operator can easily extend several solutions after reassembling between them repeatedly. In the case of arithmetic crossover, it computes the average of all the possible parental solutions. For example, in the case of two parental component mating parameters $(2, 5, 3)$ and $(4, 3, 5)$, the offspring outcomes will be $(3, 4, 4)$. This crossover operator can be extended to more than two parents. However, for the potential solution of GA, the crossover rate is considered at 0.5 as a fixed value.

### 4.2.5.2.2 Fitness

During fitness computation, the phenotype is used to evaluate the fitness function. The fitness function measures the quality of the solutions produced by GA [28]. In the optimization technique, the construction of the fitness function is one of the important parts of the mathematical modeling process. The fitness function's design decisions may be influenced by the practitioner, who can then direct the suitable searching techniques. For any instance, the fitness of impractical solutions may degrade, as in the case of penalty functions. The fitness function values of each individual objective can be combined and the weighted sum calculated when multiple objective functions need to be maximized. This method and additional techniques are used to manage multiple objective functions simultaneously.

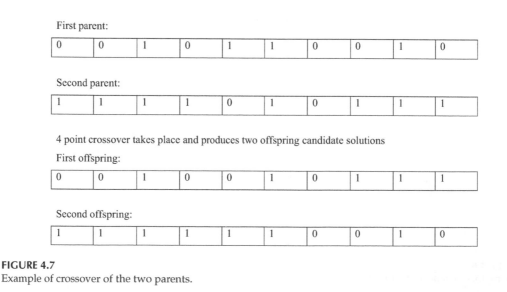

**FIGURE 4.7**
Example of crossover of the two parents.

Solution quality should be crucial for evaluation of an objective function. Though it may seem obvious, extra research is frequently necessary. A poorer solution should logically employ a worse fitness function value. Should a constrained solution that is very near to the global optimum be less suited than a mediocre solution that is workable? Should a solution in a multi-objective optimization problem perform worse in terms of fitness function value than one that is less near to the first optimum but considerably closer to the second objective's optimum, which is much more important? Selecting the optimal weights for multi-objective optimization is the correct response to all of these queries.

The main approaches are to strive to reduce the number of call of fitness function. The effectiveness of a genetic algorithm for solving problems is typically assessed in terms of how many fitness function evaluations are necessary before the optimal solution is discovered with the desired precision. Keeping the number of fitness function calls to a minimum is crucial because they might take a long time and be expensive. An excellent illustration of a rather lengthy fitness evaluation is the machine learning pipeline that was developed using a genetic algorithm. Prior to each evaluation, the machine learning pipeline must be trained on the provided dataset. To avoid over fitting, cross-validation training must be performed repeatedly, which lengthens the process. It is necessary to evaluate the prediction model's accuracy on a test set in order to acquire a precision score that can be used as the fitness function value.

### 4.2.5.2.3 Mutation

A solution is altered through mutation operators by upsetting them. The basis of mutation is random alterations [29]. The mutation rate refers to the intensity of this disturbance. The mutation rate is sometimes referred to as the step size in continuous solution spaces. There are three main criteria for mutation operators. Reachability is the first prerequisite. Every point in the solution space must be accessible from any other point there. Every area of the solution space must have a reasonable possibility of being reached. If not, there is a low likelihood that the best solution will be discovered. Not all mutation operators can ensure this property; decoder techniques, for instance, struggle to cover the entire solution space. Lack of bias is the second need for mutation operators. Without plateaus in unconstrained solution spaces, the mutation operator shouldn't cause a search to veer in a specific direction. Bias may be helpful in cases of constrained solution. Additionally, the notion of novelty search, which seeks to search in unknown regions of the solution space, introduces a bias against the mutation operator.

Scalability serves as the mutation operators' third design tenet. Each operator should provide for as much flexibility as is necessary to alter its strength. This is often possible with the majority of mutation operators that are dependent on a probability distribution. The standard deviation can scale the samples collected at random from throughout this solution space, for example in the case of the Gaussian mutation, which is based on the Gaussian distribution. Based on the chosen representation, the mutation operators are implemented. Bit-flip mutation for bit strings is frequently used. Bit-flip mutations change a zero to a one bit and vice versa with a predetermined frequency; this probability is what determines the mutation rate. It is typically selected based on how long the representation is. Every bit is reversed using the mutation rate $1/N$ for $N$ bit string.

### 4.2.5.2.4 Selection

The best offspring solutions must be chosen to be parents in the new parental population in order to enable convergence toward optimal solutions. The best offspring solutions are chosen from a surplus that is produced in order to move closer to the ideal. The population's fitness values are the foundation of this selection procedure [30, 31]. Low fitness values are preferable in minimization tasks, and vice versa in maximization tasks. With negation, a minimization task can quickly become a maximization task. Elitist selection operators choose the finest offspring solutions as parents. According to the selection process, GA researchers have used two different selection parameters, as follows: comma selection selects the $\mu$ best solutions from $\lambda$ offspring; and the plus selection process selects the $\mu$ best solutions from $\lambda$ offspring and $\mu$ parents. Except of selection parameters another three different selection algorithms used in research whose operation based on randomness.

The **roulette wheel**, sometimes referred to as fitness proportionate selection, makes uniformly distributed random selections of parental solutions. A solution's fitness determines its likelihood of selection. This fitness percentage can be thought of as the likelihood that a particular solution will be chosen. The benefit of fitness-proportional selection operators is that any solution has a chance of being chosen in the positive probability.

Another well-known selection method is **tournament selection**, which involves randomly choosing a subset of solutions from which the best ones are ultimately chosen to become new parents. Even if a solution has lower fitness values than other solutions, tournament selection offers a chance for it to prevail. Survival selection is the process of employing selection as a method to determine the parents of the next generation. Which solutions survive and which ones perish are determined by the selection operator. This viewpoint embodies Darwin's maxim of the fittest winning out in nature.

### 4.2.5.2.5 Termination

When the main evolutionary loop ends is determined by the termination condition. The genetic algorithm frequently runs for a specified number of generations. This is plausible in a variety of experimental contexts. The length of the optimization process might be limited by the time and expense of fitness function assessments. Convergence of the optimization process is another helpful termination condition. The progress of fitness function gains may drastically slow down while approaching the optimum. The evolutionary process comes to an end if no discernible process is seen. In problems involving continuous optimization, there are two alternative optima conditions.

Missing the global optimum is indicated when local optima became trapped in a condition of stagnation. In such cases the approach should be run the program with a different number of generations. In the second scenario, it is improbable that a better local optimum would be discovered if the genetic algorithm consistently accesses the same region of the optimal solutions while starting from various regions. This raises the prospect that the local optimum will be a powerful draw. Alternatively, the local optimum may be the overall one.

### 4.2.5.3 Experiments

Since the commencement of the study of genetic algorithms, experimentation has been the primary analytical instrument. Consequently, well-designed tests are crucial. The creation of a research question comes before the experimental analysis as the first task. Because the results of genetic algorithm studies are random, each run will provide a unique set of results. So the researcher needs to selects the best results. There may be one run that produces a poor result that didn't reach the optimum position, thus disturbing the average statistical output. To produce the perfect statistical output, any stochastic algorithms should to be run 25 times, where 50 or 100 is more recommendable. 15, 10, or even 5 runs may be essential as a concession in the most severe situation of phenomenally costly optimization runs.

### 4.2.5.4 Parameter Tuning Technique in Genetic Algorithm

The selection of proper parameters has a substantial impact on the effectiveness of genetic algorithm optimization procedures. The issue of how to choose the best parameter options is the biggest challenge for the researcher. Additionally, certain parameter setting and tuning activities end up being dynamic optimization problems since the best option changes as the optimization process progresses. There are taxonomies that distinguish between exogenous and endogenous variables. Exogenous parameters are general genetic algorithm parameters that define universal characteristics like population sizes and selection pressure, and chromosome-related properties are defined by endogenous parameters [31, 32].

Before GA is run, there needs to be control and tuning of the parameters. Control techniques are made to help algorithms locate the right parameters while they run. The parameters are controlled using dynamic control techniques based on static systems like the number of generations. Rechenberg's mutation rate control is one example of an adaptive parameter control strategy that makes use of feedback from the search. The automatic regulation of parameters based on a secondary genetic optimization process is known as self-adaptation. The majority of parameter tweaking and control techniques have wide application. They just require a few modest adjustments to be applied to the majority of genetic algorithms.

### 4.2.5.5 Strategy Parameters

Numerous variation operators and selection strategies have been covered in previous sections. These choices come with two disadvantages. With the supposition that you have sufficient information of the solution landscape, they provide you the chance to acquire better solutions [33–35]. Darwin's natural selection concept is incorporated into a straightforward optimization and learning system to achieve the reliable and ubiquitous solution of any complex problem. But in the real world, there are no free meals. We now address the two problems of finding the optimal solution to the original problem and of determining the optimal operators and their optimal parameters, denoted by the strategy parameter. Additionally, there are two categories of elements that might potentially impact the optimization results: global factors and local factors, which are illustrated in Figure 4.8.

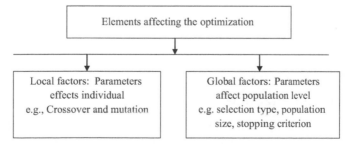

**FIGURE 4.8**
Types of factors in optimization techniques.

**TABLE 4.1**

Survey on Parameters Setting

| Si No. | References | Methodology or Parameters Settings |
|---|---|---|
| 1 | Nejati et al. [36] | Suggests ideal approach of the theoretical analysis although all the strong assumptions are hard to satisfy for the real time problem |
| 2 | De Jong [37] | Proposes that pop size = 50, $p_c$ = 0.60, $p_m$ = 0.001, G = 1, and the appropriate strategic parameters for his test functions are elitist. |
| 3 | Grefenstette [38] | To regulate the parameters of the technique in the optimization model, the author employed a meta-level GA. The best objective value obtains for the following strategy parameters: pop size = 30(80), $p_c$ = 0.95, $p_m$ = 0.01, G = 1, and elitism. |

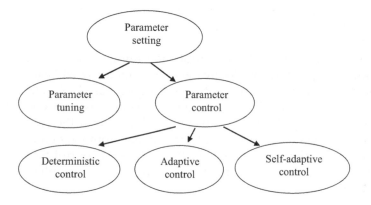

**FIGURE 4.9**
Strategy of parameter setting.

Table 4.1 describes how several researchers discussed the best ways to make optimization algorithms work better.

According to the taxonomy [39] of strategy parameter setting, parameter control can be classified into three groups as shown in Figure 4.9.

1. Deterministic control. Heuristic rules used to alter strategy parameters often solely depend on generation.
2. Adaptive control. Heuristic rules that are based on feedback from the current or prior population are used to modify the parameters of the strategy.
3. Self-adaptive control. Heuristic rules based on parameters are encoded into chromosomes.

## 4.3 Constrained Optimization

When solving constrained optimization issues in the actual world, we only looked at the area bounded by the upper and lower bounds of the variables [40–42]. When employing EAs in constraint optimization, it is imperative to address how to assess a solution that violates some constraints. Typically, we aim that all constraint should be satisfied in the final outcomes of our EAs. However, it is exceedingly inefficient to discard the violated constraint. Constrained optimization problems (COPs) can be described in following way:

Let min $p(x)$, $x \in R^n$
Such that $q_i(x) \leq 0$, $i = 1, 2 \ldots m$

$$r_j(x) = 0 \ j = n + 1, n + 2 \ldots k$$

$$L_l \leq x_l \leq U_l, \ l = 1, 2, 3 \ldots n$$

Where $L_l$ and $U_l$ are the lower and upper bounds of variable $x_l$, respectively, which form the *search space S*. $q$ inequality constraints (linear or nonlinear) and $k - q$ equality constraints (linear or nonlinear) need to be satisfied.

## 4.4 Multimodal Optimization

Sometimes during running a GA for a given problem several times, the algorithm might provide different solutions at different times. The balance between selection pressure and population variety, which is a perennial topic in constructing and analyzing Gas, can be adjusted with the use of multimodal optimization approaches. When there are numerous local optimal solutions in a solution space, the term "multimodality" is used to characterize the situation [43–45]. In this case, the algorithm identifies the one and only global optimal solution as quickly as feasible. The unconstrained maximum problem, whose target values are all positive integers, is typically used in multimodal problems.

Max $f(x) \geq 0$, $x \in R^n$

Multimodal issues might signify a variety of meanings. Other circumstances can also cause worry about coming up with multiple solutions. We are looking for all global optimal solutions, which there are plenty of. The following is a succinct explanation of the several global optimum solutions and other interesting local optimal solutions found:

- Some factors, such as dependability, manufacturing complexity, maintenance complexity, and so on will be problematic to adequately describe in implementations. Finding multiple solutions with equivalent quality provides the judgment a number of options from which to pick, based on other foggy factors.
- Finding an effective solution and doing a sensitivity analysis of a problem both benefit from having several effective alternatives which can be designed.
- Reversing the effects of genetic drift necessitates a complex trade-off between population variability and selective influence.
- In the event that the algorithm's search skills are insufficient to guarantee the discovery of the global optimum solution, the ability to locate several solutions of equivalent high quality enhances the possibility of getting the global optimal solution.

All in all, multimodal EAs must locate and keep up numerous (global/local) optimal solutions within multimodal domains.

## 4.5 Multiobjective Optimization

EAs are developed to address issues that arise in the real world, such as designing and scheduling. There are various requirements that must be met in real-world situations. Because it is challenging to compare two objectives at the same time, we occasionally wanted to express them as limitations [46–49]. We can categorize the relationship between two vectors into three groups based on Pareto's concept of dominance: one is better than other, the opposite is true, or they are not comparable. GAs are used in multiobjective optimization to address such issues. This is an exciting and popular scientific field. Let's first define the issue before providing the meanings of the terminologies utilized. Think about the next optimization challenge:

$$\min\left\{ z_1 = p_1(x), z_2 = p_2(x), \ldots\ldots\ldots\ldots z_m = p_m(x) \right\} \text{ for } x \in R^n$$

such that $q_i(x) \leq 0, i = 1, 2 \ldots\ldots\ldots\ldots\ldots.k$

$$r_j(x) = 0 \quad j = q+1, q+2 \ldots\ldots\ldots.l$$

Where $x$ is the decision variable, $p_i$ is objective $i$, $q_i$ is inequality constraint $i$, and $r_j$ is equality constraint $j$.

## 4.6 Combinatorial Optimization

The parameter optimization, or how to determine the best values for variables to obtain the maximum/minimum value of the objective function, is covered in earlier chapters. Not all problems in the real world are like this [50, 51]. It is frequently necessary to pick a few elements from a collection and arrange the constraints in such a way that the objective function has the maximum/minimal value. These issues fall under the category of combinatorial optimization.

### 4.6.1 Differential Evolution

Storn and Price first presented the stochastic, population-based optimization approach in 1996 to address nonlinear optimization issues [52–54]. The differential evolution (DE) methodology is a parallel direct search technique that makes use of NP D-dimensional parameter vectors with the values $x_{i, G}$ $i = 1; 2; \ldots;$ NP, where $G$ stands for each population generation. The initial vector population is arbitrarily chosen and should encompass literally the entire parameter space. All random decisions are picked from a uniform probability distribution. If a tentative resolution is known, the first population might be produced by adding pseudorandom variations that are allocated regularly to the nominal solution $x_{nom,0}$

Figure 4.10 shows the overall flow diagram of DE where combining the weighted difference between two population vectors with a third vector, or mutation, creates new parameter vectors. The parameters of the target vector, a different preset vector, are combined with the characteristics of the altered vector to create the so-called trial vector. Crossover is

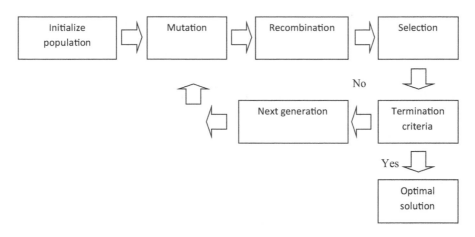

**FIGURE 4.10**
Flow diagram of differential evolution.

a term used frequently to describe parameter mixing. If the trial vector yields a lower cost function value than the target vector in the following generation, it will replace the target vector. Selection is the name of the last operation. In NP contests to occur within a single generation, every population vector may only serve as the target vector once.

### 4.6.1.1 Suitability of DE in the Field of Optimization

Users typically demand that a workable minimization strategy meet the following four criteria:

- The ability to manage cost functions that are nondifferentiable, multimodal, and nonlinear.
- A limited number of controls to direct the minimization. These factors must be reliable and simple to select.
- Recurring independent trials that consistently converge to the global minimum; good convergence qualities.

## References

1. Wein F, Dunning PD, Norato JA. A review on feature-mapping methods for structural optimization. *Structural and Multidisciplinary Optimization* 2020;62(4): 1597–1638.
2. Kochenderfer MJ, Wheeler TA. *Algorithms for optimization*. Mit Press;2019.
3. Frazier PI. A tutorial on Bayesian optimization. 2018. arXiv preprint arXiv:1807.02811.
4. Sun S, Cao Z, Zhu H, Zhao J. A survey of optimization methods from a machine learning perspective. *IEEE Transactions on Cybernetics* 2019;50(8): 3668–3681.
5. Abualigah L, Diabat A, Mirjalili S, Abd Elaziz M, Gandomi AH. The arithmetic optimization algorithm. *Computer Methods in Applied Mechanics and Engineering* 2021;376: 113609.
6. Rao RV, Pawar RB. Constrained design optimization of selected mechanical system components using Rao algorithms. *Applied Soft Computing* 2020;89: 106141.

7. Raji AK, Luta DN. Modeling and optimization of a community microgrid components. *Energy Procedia* 2019;156: 406–411.

8. Cicconi P, Nardelli M, Raffaeli R, Germani M. Integrating a constraint-based optimization approach into the design of oil and gas structures. *Advanced Engineering Informatics* 2020;45: 101129.

9. Kubica BJ. *Interval methods for solving nonlinear constraint satisfaction, optimization and similar problems.* Springer International Publishing;2019.

10. Li J, Su H. Nested primal–dual gradient algorithms for distributed constraint-coupled optimization. 2022. arXiv:2205.11119.

11. Tang J, Liu G, Pan Q. A review on representative swarm intelligence algorithms for solving optimization problems: Applications and trends. *IEEE/CAA Journal of Automatica Sinica* 2021;8(10): 1627–1643.

12. Liu Q, Li X, Liu H, Guo Z. Multi-objective metaheuristics for discrete optimization problems: A review of the state-of-the-art. *Applied Soft Computing* 2020;93: 106382.

13. Wang F, Zhang H, Zhou A. A particle swarm optimization algorithm for mixed-variable optimization problems. *Swarm and Evolutionary Computation* 2021;60: 100808.

14. Vergeynst J, Vanwyck T, Baeyens R, De Mulder T, Nopens I, Mouton A, Pauwels I. Acoustic positioning in a reflective environment: Going beyond point-by-point algorithms. *Animal Biotelemetry* 2020;8(1): 1–17.

15. Bullo F, Cisneros-Velarde P, Davydov A, Jafarpour S. From contraction theory to fixed point algorithms on Riemannian and non-Euclidean spaces. In *2021 60th IEEE Conference on Decision and Control (CDC).* IEEE;2021, December, pp. 2923–2928.

16. van den Burg GJ, Williams CK. An evaluation of change point detection algorithms. 2020 arXiv:2003.06222.

17. Abualigah L, Diabat A, Geem ZW. A comprehensive survey of the harmony search algorithm in clustering applications. *Applied Sciences* 2020;10(11): 3827.

18. Abualigah L, Elaziz MA, Hussien AG, Alsalibi B, Jalali SMJ, Gandomi AH. Lightning search algorithm: A comprehensive survey. *Applied Intelligence* 2021;51(4): 2353–2376.

19. Rashedi E, Rashedi E, Nezamabadi-Pour H. A comprehensive survey on gravitational search algorithm. *Swarm and Evolutionary Computation* 2018;41: 141–158.

20. Amine K. Multiobjective simulated annealing: Principles and algorithm variants. *Advances in Operations Research* 2019;2019: 8134674.

21. Delahaye D, Chaimatanan S, Mongeau M. Simulated annealing: From basics to applications. In *Handbook of metaheuristics.* Springer;2019, pp. 1–35.

22. Abdel-Basset M, Ding W, El-Shahat D. A hybrid Harris Hawks optimization algorithm with simulated annealing for feature selection. *Artificial Intelligence Review* 2021;54(1): 593–637.

23. Liang H, Zou J, Zuo K, Khan MJ. An improved genetic algorithm optimization fuzzy controller applied to the wellhead back pressure control system. *Mechanical Systems and Signal Processing* 2020;142: 106708.

24. Mayer MJ, Szilágyi A, Gróf G. Environmental and economic multi-objective optimization of a household level hybrid renewable energy system by genetic algorithm. *Applied Energy* 2020;269: 115058.

25. Ding X, Zheng M, ZhengX. The application of genetic algorithm in land use optimization research: A review. *Land* 2021;10(5): 526.

26. Hassanat A, Almohammadi K, Alkafaween EA, Abunawas E, Hammouri A, Prasath VS. Choosing mutation and crossover ratios for genetic algorithms—A review with a new dynamic approach. *Information* 2019;10(12): 390.

27. Stemper L, Tunes MA, Tosone R, Uggowitzer PJ, Pogatscher S. On the potential of aluminum crossover alloys. *Progress in Materials Science* 2022;124: 100873.

28. Kumar A, Sinha N, Bhardwaj A. A novel fitness function in genetic programming for medical data classification. *Journal of Biomedical Informatics* 2020;112: 103623.

29. Stenson PD, Mort M, Ball EV, Chapman M, Evans K, Azevedo L, … Cooper DN. The Human Gene Mutation Database (HGMD®): Optimizing its use in a clinical diagnostic or research setting. *Human Genetics* 2020;139(10): 1197–1207.

30. Khaire UM, Dhanalakshmi R. Stability of feature selection algorithm: A review. *Journal of 1King Saud University—Computer and Information Sciences* 2019;34: 1060–1073.

31. Hassanat A, Almohammadi K, Alkafaween EA, Abunawas E, Hammouri A, Prasath VS. Choosing mutation and crossover ratios for genetic algorithms—A review with a new dynamic approach. *Information* 201910(12): 390.

32. Bora TC, Mariani VC, dos Santos Coelho L. Multi-objective optimization of the environmental-economic dispatch with reinforcement learning based on non-dominated sorting genetic algorithm. *Applied Thermal Engineering* 2019;146: 688–700.

33. Eltamaly AM. A novel strategy for optimal PSO control parameters determination for PV energy systems. *Sustainability* 2021;13(2): 1008.

34. Vyacheslavovich BH, Viktorovna BO. Optimizing strategy parameters" on-condition" maintenance with constant monitoring frequency. In *The 1 st International Scientific and Practical Conference "Eurasian Scientific Discussions"(February 13–15, 2022)* Barca Academy Publishing, Barcelona, Spain; 2022, February, 582 p. (p. 136).

35. Mostafa M, Rezk H, Aly M, Ahmed EM. A new strategy based on slime mould algorithm to extract the optimal model parameters of solar PV panel. *Sustainable Energy Technologies and Assessments* 2020;42: 100849.

36. Nejati SA, Chong B, Alinejad M, Abbasi S Optimal scheduling of electric vehicles charging and discharging in a smart parking-lot. In *2021 56th International Universities Power Engineering Conference (UPEC)*. IEEE, pp. 1–6.

37. De Jong KLearning with genetic algorithms: An overview. *Machine Learning* 1988;3(2): 121–138.

38. Grefenstette JJ (Ed.). *Genetic algorithms and their applications: Proceedings of the second international conference on genetic algorithms*. Psychology Press;2013.

39. Krink T, Ursem RK. Parameter control using the agent based patchwork model. In *Proceedings of the 2000 Congress on Evolutionary Computation. CEC00 (Cat. No. 00TH8512)*. IEEE;2000, July, Vol. 1, pp. 77–83.

40. Bertsekas DP. *Constrained optimization and Lagrange multiplier methods*. Academic Press;2014.

41. Box MJ. A new method of constrained optimization and a comparison with other methods. *The Computer Journal* 1965;8(1): 42–52.

42. Homaifar A, Qi CX, Lai SH. Constrained optimization via genetic algorithms. *Simulation* 1994;62(4): 242–253.

43. Yang XS. Firefly algorithms for multimodal optimization. In *International Symposium on Stochastic Algorithms*. Springer, Berlin, Heidelberg;2009, October, pp. 169–178.

44. Das S, Maity S, Qu BY, Suganthan PN. Real-parameter evolutionary multimodal optimization—A survey of the state-of-the-art. *Swarm and Evolutionary Computation* 2011;1(2): 71–88.

45. Qu BY, Suganthan PN, Liang JJ. Differential evolution with neighborhood mutation for multimodal optimization. *IEEE Transactions on Evolutionary Computation* 2012;16(5): 601–614.

46. Miettinen K. *Nonlinear multiobjective optimization*. Springer Science and Business Media;2012, Vol. 12.

47. Sawaragi Y, Nakayama H, Tanino T (Eds.). *Theory of multiobjective optimization*. Elsevier;1985.

48. Abraham A, Jain L. *Evolutionary multiobjective optimization*. Springer;2005, pp. 1–6.

49. Collette Y, Siarry P. *Multiobjective optimization: Principles and case studies*. Springer Science and Business Media;2004.

50. Korte BH, Vygen J, Korte B, Vygen J. *Combinatorial optimization*. Springer;2011, Vol. 1).

51. Cook W, Lovász L, Seymour PD (Eds.). *Combinatorial optimization: Papers from the DIMACS special year*. American Mathematical Soc;1995, Vol. 20.

52. Price KV. Differential evolution. In *Handbook of optimization*. Springer;2013, pp. 187–214.

53. Feoktistov V. *Differential evolution*. Springer;2006, pp. 1–24.

54. Fleetwood K. An introduction to differential evolution. In *Proceedings of Mathematics and Statistics of Complex Systems (MASCOS) One Day Symposium*, 26th November, Brisbane, Australia;2004, November, pp. 785–791.

# 5

## Modeling of ANFIS (Adaptive Fuzzy Inference System) System

### 5.1 Introduction

Soft computing techniques like neural networks (NN), fuzzy logic (FL), and genetic algorithms (GA) draw their inspiration from biological computational processes and nature's approaches to problem-solving [1–3]. NNs are simplified representations of the human nervous system that imitate our capacity for situational adaptation and experience-based learning. Chapter 3 of the book discusses three NN systems, each representing the three major classes of NN architectures, namely single layer feedforward, multilayer feedforward, and recurrent network architectures. The backpropagation network is a multilayer feedforward network architecture with gradient descent learning. Associative memories are single layer feedforward or recurrent network architectures adopting Hebbian learning [4, 5]. ART networks are recurrent network architectures with a kind of competitive learning termed adaptive resonance theory. Systems that use fuzzy sets' imprecision or ambiguity in their input and output descriptions are addressed by fuzzy logic systems. Fuzzy sets have no clearly defined boundaries and offer a gradual change in an element's membership or absence from the collection [6]. In Chapter 2 of the book, fuzzy logic's foundational ideas and applications are covered. Genetic algorithms (GAs) inspired by the process of biological evolution, are adaptive search and optimization algorithms. Chapter 4 of the book discusses the basic concepts, namely the genetic inheritance operators and applications of GA. Each technology has successfully solved a variety of issues originating from various domains on its own terms and merit [7]. At the same time, as mentioned in Chapter 1, various attempts have been successfully made to synergize the three different technologies in whole or in part, to solve problems for which these technologies could not find solutions individually. By properly integrating them, synergy or hybridization aims to overcome the drawbacks of one technology's application while utilizing the advantages of the other. When one technology applied alone has been unable to produce an effective solution, the complexity of the problem has more typically called for a careful mix of the technologies [8–10].

DOI: 10.1201/9781003216001-6

## 5.2 Hybrid Systems

Hybrid systems are ones that combine more than one technique to address the issue. Hybrid systems can be classified into the following categories [11]:

1. Sequential hybrids
2. Auxiliary hybrids
3. Embedded hybrids

### 5.2.1 Sequential Hybrid Systems

Sequential hybrid systems are similar to a pipeline technology [12, 13]. Thus, the output of one technology becomes the input of another. One of the weakest types of hybridization is sequential hybrid systems, which lack an integrated fusion of the technologies. For an example, optimal parameters of a given problem are preprocess by GA, and it further proceeds to the NN model for further processing.

### 5.2.2 Auxiliary Hybrid Systems

In this scenario, one technology uses the other as a "subroutine" or changes the data according to their requirement [14–16]. The first technology's information is processed by the second technology before being passed on for additional usage. Despite being superior to sequential hybrids, this sort of hybridization is only considered to be of an intermediate degree. As an illustration, consider a neurogenetic system where a NN uses a GA to optimize the parameters that define its structural architecture performance.

### 5.2.3 Embedded Hybrid Systems

The involved technologies in embedded hybrid systems are integrated to the point that they appear to be entangled [17–19]. It would seem that no technology can be employed to solve the problem without the others because the fusion is so complete. The hybridization is complete in this case. An NN, for instance, may be part of a NN–FL hybrid system that processes fuzzy inputs and also extracts fuzzy outputs.

## 5.3 Neuro-Fuzzy Hybrids

One of the hybrid system types that have received the most attention, this one has produced an incredible number of papers and research findings [20]. Fuzzy logic and neural networks are two different approaches to dealing with uncertainty. Each of them has strengths and weaknesses of its own. Neural networks are well suited for classifying phenomena into specified groups because they can model complex nonlinear interactions [21]. On the other hand, output precision is frequently constrained and only permits minimization of least squares errors, not zero error. Additionally, a NN's training time need can be rather high [24]. Additionally, the training data must be carefully picked to span the whole range over which the various variables are anticipated to change [22, 23].

Two perspectives can be taken on this hybridization. One is to add fuzzy functionality to NNs to increase expressiveness and flexibility of a network in uncertain circumstances.

The second component involves giving fuzzy systems access to neural learning capabilities so they may become more environment-adaptive. This method is sometimes referred to as "NN driven fuzzy reasoning" in the literature [25, 26].

### 5.3.1 Adaptive Neuro-Fuzzy Interference System (ANFIS)

ANFIS combines the ANN and fuzzy logic soft computing techniques [27]. The qualitative ideas of human knowledge and insights into the methodology of exact quantitative analysis can be altered by fuzzy logic. However, it lacks a clearly defined mechanism for converting human cognition into a rule-based FIS, and it also updates the MFs [28]. Compared to ANN, it has a higher capacity for learning to adapt to its surroundings. As a result, while determining rules, the ANN can be utilized to automatically adjust the MFs and reduce the rate of error [29].

#### 5.3.1.1 Fuzzy Inference System (FIS)

A FIS was created using three main parts: basic rules, which are used to choose fuzzy logic "if-then" rules based on fuzzy set membership, and fuzzy inference techniques (FIS) to reason from basic rules to produce the output. The FIS's intricate construction is depicted in Figure 5.1. When the input containing the real value is fuzzifying by its membership function, the input will function as the fuzzy value system (FIS), which is range between 0 and 1 [30]. The knowledge base refers to the fundamental laws and databases, both of which are crucial components in decision-making. A fuzzy database typically contains definitions, such as details on fuzzy set parameters with defined functions. In order to create a database, a universe must normally be defined, the number of linguistic values to be utilized for each linguistic variable must be determined, and a membership function must be established. It has fuzzy logic operators and an "if–then" conditional statement that is based on the rules. The fundamental rules can be created manually or automatically, with the searching rules employing numerical input–output data. FIS comes in a variety of forms, including Mamdani, Takagi-Sugeno, and Tsukamoto [31, 32]. It was discovered that the ANFIS method was applied frequently using the FIS of the Takagi-Sugeno model.

#### 5.3.1.2 Adaptive Network

One of the best examples of a multilayer feedforward neural network is the adaptive network shown in Figure 5.2. These networks frequently employ supervised learning

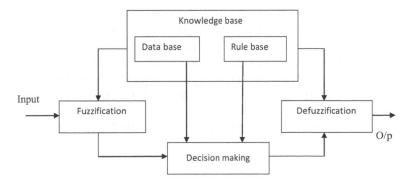

**FIGURE 5.1**
Fuzzy inference systems.

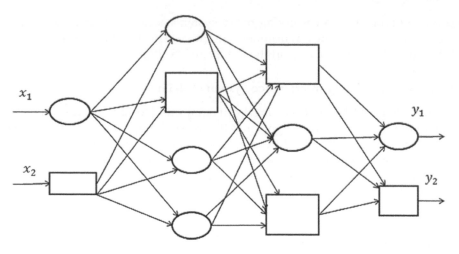

**FIGURE 5.2**
Adaptive network.

algorithms during the learning phase. Additionally, in this architecture a number of adaptive nodes are connected directly to one another without the need of weight values [33]. The output of each node depends upon the incoming signals and parameters, and each node performs a different set of tasks and has varied roles. An applied learning rule may have an impact on the node's parameters and lessen the likelihood of errors at the adaptive network's output [34]. In adaptive networks nowadays, gradient descent and back-propagation are still utilized as a learning technique. Still, the back-propagation method has been proven to have flaws, which further limit the ability and precision of adaptive networks to make judgments. Major issues with the back-propagation algorithm include its slow convergence rate and propensity to remain stuck in local minima. As a result, [35] suggested a hybrid learning algorithm as an alternate learning method. This method has a greater capacity to accelerate convergence and prevent becoming locked in local minima.

## 5.4 ANFIS Architecture

The ANFIS strategy is utilized to manage nondirect and complex problems. In ANFIS hybrid intelligent systems, a straightforward informational index produces the desired output of the fuzzy logic controller through an interconnected NN handling components by means of weighted data associations [36, 37]. ANFIS consolidates the quality of the two intelligent strategies FLC and NN into a solitary strategy. ANFIS model parameters of a FIS are tuned by the NN learning strategies. A five-layer ANFIS structure appears in Figure 5.3.

- It amends fuzzy if-then principles to depict the conduct of a nonstraight and complex framework.
- It doesn't require prior human skill.
- It is simple to actualize.
- It empowers quick and precise thinking and learning quality.

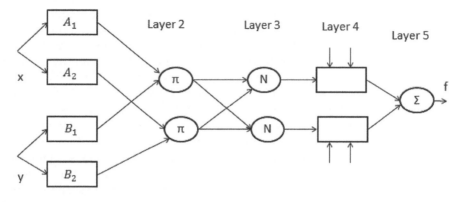

**FIGURE 5.3**
ANFIS architecture for five layers.

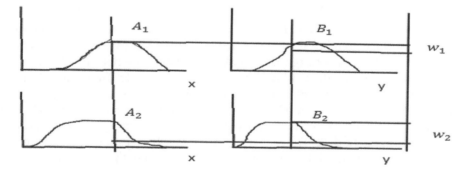

**FIGURE 5.4**
Firing of rules for different membership functions.

- Because of legitimate choice of a reasonable decision of membership function, strong speculation, and brilliant clarification of fuzzy guidelines, it offers the desired output.
- It is simple to coordinate the both etymological and numeric information for critical thinking.

Figure 5.3 demonstrates the ANFIS engineering executed by the two principles where fixed node and versatile node are spoken to by circle and square. In Figure 5.4 represent first order Sugeno model of ANFIS architecture. In Layer 1, every node is versatile and outputs of Layer 1 are the fuzzy membership of input. In Layer 2, the nodes are fixed and the fuzzy administrators fuzzify the inputs by utilizing AND operator. Symbol $\Pi$ showing a basic multiplier activity. Output of layer 2 provides the standardized firing qualities. In Layer 3, fixed nodes named as N which normalized the firing strengths from the previous layer. In Layer 4, nodes are versatile and output of every node is basically standardized firing quality and a first order polynomial. In Layer 5, it only consist a single fixed node $\Sigma$ which plays out the summation of every single approaching sign.

### 5.4.1 Hybrid Learning Algorithm

In this algorithm, the ANFIS model is a blend of least squares strategies and gradient descent strategies [38]. In the forward pass learning calculation, least square techniques are utilized to determine node outputs until Layer 4 and the consequent parameters. In the backward path, gradient descendent method sends the error signals to the previous stage and updated parameters. The hybrid learning approach is quicker than the original back-propagation technique because of diminishing search space dimensions. The ideal estimation of these parameters is controlled by the least squares technique. At the point when these parameters are variable, the search space in a hybrid learning process increases; subsequently, the convergence rate of the training datasets ought to be slower [39, 40]. To tackle the issue of search space, the hybrid algorithm of ANFIS joins two techniques: (1) least squares strategy and (2) gradient descent technique. Where the least squares strategy is used to advance the subsequent parameters and the gradient descent technique is utilized to streamline the reason parameters. From the review it has been seen that the hybrid algorithm gives the high level of proficiency in preparing the ANFIS frameworks. A different hybrid ANFIS learning process is shown in Figure 5.5.

### 5.4.2 Derivation of Fuzzy Model

As previously mentioned, the identification of an ideal fuzzy model with regard to the training data simplifies to a linear least-squares estimate problem in the ANFIS model for a given set of rules [41]. This suggests a quick and reliable method for identifying fuzzy models from input–output data. When creating a fuzzy model from data, this method chooses the crucial input variables by fusing the cluster estimation method (CEM) and the least squares estimation algorithm (LSM). There are two steps to the method [42]: (1) In the first stage, a fuzzy model is implemented from a given set of input and output data using a cluster estimation method. (2) The following phase involves testing the relevance of each variable in the first fuzzy model in order to determine the key input variables.

#### 5.4.2.1 Extracting the Initial Fuzzy Model

An initial fuzzy model must be derived in order to begin the modeling process. This model is necessary to determine the final fuzzy model's number of inputs, linguistic variables,

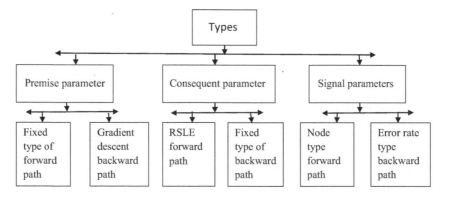

**FIGURE 5.5**
Hybrid learning process.

and consequently, rules. Before the final optimal model can be developed, the initial model must choose the model selection criteria as well as the input variables for the final model. Depending on the provided fuzzy rules, either the subtractive clustering technique or the grid partitioning method [43–45] can be used to select the initial fuzzy model.

### 5.4.2.2 Subtractive Clustering Technique

This method is used on the input–output data pairs collected from the system that has to be modeled as the first stage in obtaining the initial fuzzy model by subtractive clustering [46]. The input–output data pairs' cluster centers can be found using the cluster estimation technique. As a result, it is easier to identify the scattered rules in input–output space because each cluster center indicates the existence of a rule. Additionally, it aids in determining the parameters' values for the underlying assumption. This is significant because, throughout the neural network training session, the model will eventually quickly converge to its final value if the beginning value is close to the final value [47]. In this clustering method, the potentials of each input and output data point are determined using their Euclidian distances from one another. Cluster centers are points that have potential over a particular threshold value. The initial fuzzy model can then be recovered after the cluster centers have been determined because the centers will also indicate how many linguistic variables there are (Figure 5.6).

Let's have a look at a set of $n$ data points in an $M$-dimensional space $\{x_1, x_2, x_3 \ldots x_n\}$. In order for the data points to be circumscribed by a unit hypercube, it is assumed that they have been standardized in each dimension [48]. Every single piece of data is regarded as a possible cluster center. $P_i$ is a potential of data point $x_i$ that can be presented as

$$P_i = \sum_{i=1}^{n} e^{\|x_i - x_j\|^2} \tag{5.1}$$

Where $\alpha = \dfrac{4}{r_a^2}$,

$\|x_i - x_j\|$ is a Euclidean distance, and $r_a$ is a positive constant. If $x_1^*$ is the location and $P_1^*$ is its potential value of the first cluster center, then the revised potential formula for each data point $x_i$ is presented by

$$P_i = P_i - P_1^* e^{-\beta \|x_i - x_1^*\|^2} \tag{5.2}$$

Where $\beta = \dfrac{4}{r_b^2}$ and

$r_b$ is a positive constant. As a result, each data point has potential deducted from it based on how far it is from the cluster center. Depending on how far each data point is from the second cluster center, their potential is further diminished. In general, after the $k$th cluster center has been obtained, the potential of each data point is revised by the formula

$$P_i = P_i - P_1^* e^{-\beta \|x_i - x_k^*\|^2} \tag{5.3}$$

Where $x_k^*$ is the location of the $k$th cluster center and $P_k^*$ is the potential value.

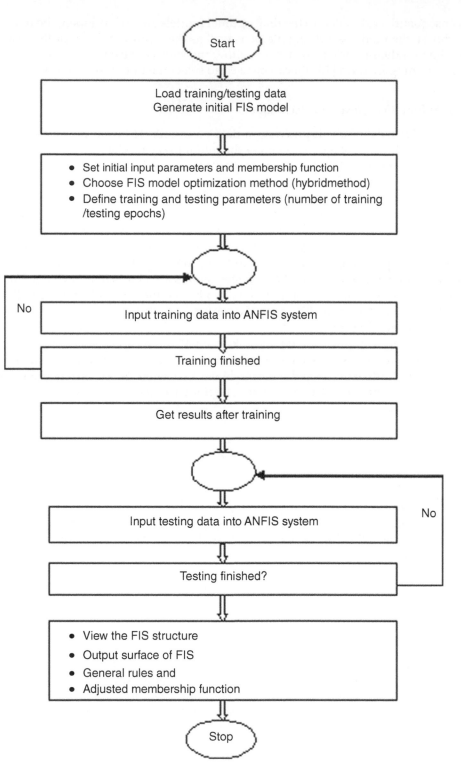

**FIGURE 5.6**
Flowchart of ANFIS model training process [36].

The process is acquiring a new cluster center and revising potentials repeats until the stopping criterion $P_k^* < 0.15\ P_1^*$ [49] is satisfied. It is considered that $\{x_1^*,\ x_2^*,\ \ldots\ x_c^*\ \}$ is a set of c cluster centers in an $M$ dimensional space. Cluster arranged in such a way that if first $N$ dimensions present input variables, then the last $M-N$ dimensions correspond to output variables. Each vector $x_i^*$ is divided into two components $y_i^*$ and $z_i^*$, where $y_i^*$ is the location of the center of the input cluster and $z_i^*$ is the location of the center of the output cluster. Therefore $x_i^*$ may be represented as $x_i^* = [y_i^*;\ z_i^*]$. The degree of fulfillment of the rule can be defined as

$$\mu_i = e^{-\alpha\left\|y-y_i^*\right\|^2} \tag{5.4}$$

Where $\alpha$ is a constant defined by Eq. (5.5). Output vector $z$ is computed as

$$z = \frac{\displaystyle\sum_{i=1}^{c} \mu_i z_i^*}{\displaystyle\sum_{i=1}^{c} \mu_i} \tag{5.5}$$

### 5.4.2.3  Grid Partitioning Technique

The second technique for defining the first fuzzy model's rules is grid partitioning [50–52]. When there are fewer inputs and membership functions, this strategy is employed. To build the fuzzy rules' antecedents in this situation, the input spaces are separated into a number of fuzzy regions. The fuzzy space for a two-input model with three membership functions for each input is shown in grid-partitioned form in Figure 5.7. The ordinate and abscissa of the input space are represented by the two dimensions. The rules obtained using one of the two ways is then improved using Jang's ANFIS methodology [36].

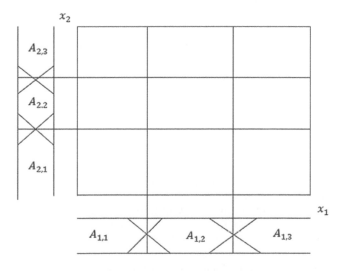

**FIGURE 5.7**
Grid search partitioning techniques for two input variables with three membership functions.

### 5.4.2.4 C-Mean Clustering

When a dataset is clustered, it is divided into groups, with similar datasets belonging to one cluster while dissimilar datasets are assigned to another [53]. In order to automatically find a flaw, fuzzy C-means (FCM) clustering was employed in this study. The FCM algorithm, which is an unsupervised algorithm, is one of the most used fuzzy clustering techniques. A nonmonitoring clustering method is the FCM clustering. The FCM, an enhancement and modification of K-means clustering, uses a dataset of $x_i$ data points to create C clusters by minimizing the objective function U. The proposed method reduces the number of membership functions and rules by using FCM clustering [54]. The purpose of the FCM clustering algorithm is very similar to the $k$-means algorithm, and its definition is given in Eq. (5.7):

$$\sum_{j=1}^{k}\sum_{x_i \in c_k} u_{ij}^m \left( x_i - \mu_j \right)^2 \tag{5.6}$$

Where $u_{ij}$ is the degree observation, $x_i$ belongs to a cluster $C_j$, and $\mu_j$ is the center of the cluster $j$. The variable $u_{ij}^m$ is defined as Eq. (5.7):

$$u_{ij}^m = \frac{1}{\displaystyle\sum_{i=1}^{k}\left( \frac{\left[ x_{i-C_j} \right]}{\left[ x_{i-C_k} \right]} \right)^{\frac{2}{m-1}}} \tag{5.7}$$

The parameter $m$ determines the degree of cluster fuzziness and is a real value greater than 1 ($1 \leq m < \infty$). The value of $m$ near to 1 results in a cluster solution that is closer to the solution of hard clustering, such $k$-means, while a value of $m$ close to infinite results in total fuzziness. The centroid of a cluster in fuzzy clustering is the average of all points, weighted according to how much they belong to the cluster shown in Eq. (5.8):

$$C_j = \frac{\displaystyle\sum_{x \in C_j} u_{ij}^m . x}{\displaystyle\sum_{x \in C_j} u_{ij}^m} \tag{5.8}$$

## References

1. Rajasekaran S, Pai GV. *Neural networks, fuzzy logic and genetic algorithm: Synthesis and applications (with CD)*. PHI Learning Pvt. Ltd;2003.
2. Ansari A, Bakar AA. A comparative study of three artificial intelligence techniques: Genetic algorithm, neural network, and fuzzy logic, on scheduling problem. *2014 4th International Conference on Artificial Intelligence with Applications in Engineering and Technology*. IEEE;2014, December, pp. 31–36.
3. Yardimci A. Soft computing in medicine. *Applied Soft Computing* 2009;9(3): 1029–1043.
4. Hanne T, Dornberger R. Computational intelligence. In *Computational intelligence in logistics and supply chain management*. Springer;2017, pp. 13–41.

5. Hanafy TO, Zaini H, Shoush KA, Aly AA. Recent trends in soft computing techniques for solving real time engineering problems. *International Journal of Control, Automation and Systems* 2014;3(1): 27–33.

6. Huang Y, Lan Y, Thomson SJ, Fang A, Hoffmann WC, Lacey RE. Development of soft computing and applications in agricultural and biological engineering. *Computers and Electronics in Agriculture* 2010;71(2): 107–127.

7. Das SK, Kumar A, Das B, Burnwal AP. On soft computing techniques in various areas. *Computer Science Information Technology* 2013;3(59): 166.

8. Rao KK, Svp Raju G. An overview on soft computing techniques. *International Conference on High Performance Architecture and Grid Computing*. Springer;2011, July, pp. 9–23.

9. Falcone R, Lima C, Martinelli E. Soft computing techniques in structural and earthquake engineering: A literature review. *Engineering Structures* 2020;207: 110269.

10. Siddique N, Adeli H. *Computational intelligence: Synergies of fuzzy logic, neural networks and evolutionary computing*. John Wiley and Sons;2013.

11. Gray A, Sallis P, MacDonell S. (1997). Software forensics: Extending authorship analysis techniques to computer programs.

12. Yanaka K, Nakayama K, Shinke T, Shinkura Y, Taniguchi Y, Kinutani H, … Hirata KI. Sequential hybrid therapy with pulmonary endarterectomy and additional balloon pulmonary angioplasty for chronic thromboembolic pulmonary hypertension. *Journal of the American Heart Association* 2018;7(13): e008838.

13. Pandya PK, Wang HP, Xu QW. Towards a theory of sequential hybrid programs. *Programming concepts and methods PROCOMET'98*. Springer;1998, pp. 366–384.

14. Pettersson S, Lennartson B Stability and robustness for hybrid systems. *Proceedings of 35th IEEE Conference on Decisionp and Control*. IEEE;1996, December, Vol. 2, pp. 1202–1207.

15. Erdinc O, Uzunoglu M. Recent trends in PEM fuel cell-powered hybrid systems: Investigation of application areas, design architectures and energy management approaches. *Renewable and Sustainable Energy Reviews* 2010;14(9): 2874–2884.

16. Camacho EF, Ramírez DR, Limón D, De La Peña DM, Alamo T. Model predictive control techniques for hybrid systems. *Annual Reviews in Control* 2010;34(1): 21–31.

17. Alur R. Formal verification of hybrid systems. *Proceedings of the Ninth ACM International Conference on Embedded Software*. 2011, October, pp. 273–278.

18. Branicky MS. Introduction to hybrid systems. In *Handbook of networked and embedded control systems*. Birkhäuser;2005, pp. 91–116.

19. Lygeros J. An overview of hybrid systems control. In *Handbook of Networked and Embedded Control Systems*, Lygeros J (Ed.). Springer;2005, pp. 519–537.

20. de Campos Souza PV. Fuzzy neural networks and neuro-fuzzy networks: A review the main techniques and applications used in the literature. *Applied Soft Computing* 2020;92: 106275.

21. Rutkowska D. *Neuro-fuzzy architectures and hybrid learning*. Springer Science and Business Media;2001, Vol. 85.

22. Kim J, Kasabov N. HyFIS: Adaptive neuro-fuzzy inference systems and their application to nonlinear dynamical systems. *Neural Networks* 1999;12(9): 1301–1319.

23. Kawamura K, Miyamoto A. Condition state evaluation of existing reinforced concrete bridges using neuro-fuzzy hybrid system. *Computers and Structures* 2003;81(18–19): 1931–1940.

24. Li K, Su H, Chu J. Forecasting building energy consumption using neural networks and hybrid neuro-fuzzy system: A comparative study. *Energy and Buildings* 2011;43(10): 2893–2899.

25. Porwal A, Carranza EJM, Hale M. A hybrid neuro-fuzzy model for mineral potential mapping. *Mathematical Geology* 2004;36(7): 803–826.

26. Taylan O, Karagözoğlu B. An adaptive neuro-fuzzy model for prediction of student's academic performance. *Computers and Industrial Engineering* 2009;57(3): 732–741.

27. Walia N, Singh H, Sharma A. ANFIS: Adaptive neuro-fuzzy inference system-a survey. *International Journal of Computer Applications* 2015;123(13): 32–38.

28. Al-Hmouz A, Shen J, Al-Hmouz R, Yan J. Modeling and simulation of an adaptive neuro-fuzzy inference system (ANFIS) for mobile learning. *IEEE Transactions on Learning Technologies* 2011;5(3): 226–237.

29. Çaydaş U, Hasçalık A, Ekici S. An adaptive neuro-fuzzy inference system (ANFIS) model for wire-EDM. *Expert Systems with Applications* 2009;36(3): 6135–6139.
30. Cabalar AF, Cevik A, Gokceoglu C. Some applications of adaptive neuro-fuzzy inference system (ANFIS) in geotechnical engineering. *Computers and Geotechnics* 2012;40: 14–33.
31. Sihag P, Tiwari NK, Ranjan S. Prediction of unsaturated hydraulic conductivity using adaptive neuro-fuzzy inference system (ANFIS). *ISH Journal of Hydraulic Engineering* 2019;25(2): 132–142.
32. Kakar M, Nyström H, Aarup LR, Nøttrup TJ, Olsen DR. Respiratory motion prediction by using the adaptive neuro fuzzy inference system (ANFIS). *Physics in Medicine and Biology* 2005;50(19): 4721.
33. Gross T, D'Lima CJD, Blasius B. Epidemic dynamics on an adaptive network. *Physical Review Letters* 2006;96(20): 208701.
34. Tero A, Takagi S, Saigusa T, Ito K, Bebber DP, Fricker MD, … Nakagaki T. Rules for biologically inspired adaptive network design. *Science* 2010;327(5964): 439–442.
35. Gluck MA, Bower GH. Evaluating an adaptive network model of human learning. *Journal of Memory and Language* 1988;27(2): 166–195.
36. Jang JS. ANFIS: Adaptive-network-based fuzzy inference system. *IEEE Transactions on Systems, Man, and Cybernetics* 1993;23(3): 665–685.
37. Jovanovic BB, Reljin IS, Reljin BD. Modified ANFIS architecture-improving efficiency of ANFIS technique. *7th Seminar on Neural Network Applications in Electrical Engineering, 2004. NEUREL 2004. 2004.* IEEE;2004, September, pp. 215–220.
38. Castro JR, Castillo O, Melin P, Rodríguez-Díaz A. A hybrid learning algorithm for a class of interval type-2 fuzzy neural networks. *Information Sciences* 2009;179(13): 2175–2193.
39. Davanipoor M, Zekri M, Sheikholeslam F. Fuzzy wavelet neural network with an accelerated hybrid learning algorithm. *IEEE Transactions on Fuzzy Systems* 2011;20(3): 463–470.
40. Zhang D, Zhang N, Ye N, Fang J, Han X. Hybrid learning algorithm of radial basis function networks for reliability analysis. *IEEE Transactions on Reliability* 2020;70(3): 887–900.
41. Takagi T, Sugeno M. Derivation of fuzzy control rules from human operator's control actions. *IFAC Proceedings Volumes* 1983;16(13): 55–60.
42. Tanaka K, Hori T, Wang HO. A fuzzy Lyapunov approach to fuzzy control system design. *Proceedings of the 2001 American Control Conference.(Cat. No. 01CH37148).* IEEE;2001, June, Vol. 6, pp. 4790–4795.
43. Chiu SL. Selecting input variables for fuzzy models. *Journal of Intelligent & Fuzzy Systems* 1996;4(4): 243–256.
44. Chen MY, Linkens DA. Rule-base self-generation and simplification for data-driven fuzzy models. *Fuzzy Sets and Systems* 2004;142(2): 243–265.
45. Wang H, Kwong S, Jin Y, Wei W, Man KF. Multi-objective hierarchical genetic algorithm for interpretable fuzzy rule-based knowledge extraction. *Fuzzy Sets and Systems* 2005;149(1): 149–186.
46. Hammouda K, Karray F. *A comparative study of data clustering techniques.* University of Waterloo;2000.
47. Priyono A, Ridwan M, Alias AJ, Rahmat RAO, Hassan A, Ali MAM. Generation of fuzzy rules with subtractive clustering. *Journal Technology* 2005;43: 143–153.
48. Chen JY, Qin Z, Jia J. A PSO-based subtractive clustering technique for designing RBF neural networks. *2008 IEEE Congress on Evolutionary Computation (IEEE World Congress on Computational Intelligence).* IEEE;2008, June, pp. 2047–2052.
49. Pereira R, Fagundes A, Melicio R, Mendes VMF, Figueiredo J, Quadrado JC. Fuzzy subtractive clustering technique applied to demand response in a smart grid scope. *Procedia Technology* 2014;17: 478–486.
50. Pieters TA, Conner CR, Tandon N. Recursive grid partitioning on a cortical surface model: An optimized technique for the localization of implanted subdural electrodes. *Journal of Neurosurgery* 2013;118(5): 1086–1097.
51. Eldawy A, Alarabi L, Mokbel MF. Spatial partitioning techniques in SpatialHadoop. *Proceedings of the VLDB Endowment* 2015;8(12): 1602–1605.

52. Berger M, Rigoutsos I. An algorithm for point clustering and grid generation. *IEEE Transactions on Systems, Man, and Cybernetics* 1991;21(5): 1278–1286.
53. Selvakumar J, Lakshmi A, Arivoli T. Brain tumor segmentation and its area calculation in brain MR images using K-mean clustering and Fuzzy C-mean algorithm. *IEEE-International Conference On Advances In Engineering, Science And Management (ICAESM-2012)*. IEEE;2012, March, pp. 186–190.
54. Chowdhary CL, Mittal M, Pattanaik PA, Marszalek Z. An efficient segmentation and classification system in medical images using intuitionist possibilistic fuzzy C-mean clustering and fuzzy SVM algorithm. *Sensors* 2020;20(14): 3903.

# 6

# Machine Learning Techniques for Cognitive Modeling

## 6.1 Introduction

Human individuals are born with the ability to learn. As a result, people gain the ability to perform better while carrying out similar tasks [1–10]. An outline of the learning concept that may be applied to machines to enhance their performance is given in this chapter. Three main categories can be used to group machine learning: supervised learning, unsupervised learning, and reinforcement learning [2, 11–14]. A trainer is necessary for supervised learning, and they provide the input–output training examples. In order to produce the appropriate output patterns from a given input pattern, the learning system adjusts its parameters using a few methods [1–4]. Without instructors, the learner must adjust the parameters on their own because the expected outcome for a given input instance is unknown. Unsupervised learning is the name given to this sort of learning. Between supervised and unsupervised categories, there is a third category called reinforcement learning. The learner in reinforcement learning receives feedback from its surroundings even though it is not explicitly aware of the input–output instances [5, 6].

## 6.2 Classification of Machine Learning

The learner can determine whether its actions on the environment are rewarding or punitive with the use of the feedback signals. Thus, based on the states of its activities, the learner adjusts its settings. The most popular supervised learning strategies are inductive and analogical learning. Decision tree and version space–based learning are both parts of the inductive learning technique that is discussed in this chapter. Illustrational examples are used to briefly introduce analogous learning. Here, a clustering issue is used to explain the idea of unsupervised learning. Temporal difference learning (TD) and Q-learning are included in the section on reinforcement learning. Fourth category learning named as Inductive logic programming (ILP) has recently been identified in the fields of knowledge engineering. Figure 6.1 shows the basic classification of machine learning techniques.

In this chapter, the fundamentals of inductive logic programming have also been briefly covered. A brief explanation of the computational theory of learning concludes the chapter. With this theory as a foundation, it is possible to gauge how well a computer learns from training examples by counting them.

DOI: 10.1201/9781003216001-7

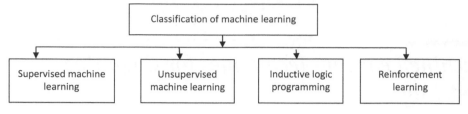

**FIGURE 6.1**
Classification of machine learning.

### 6.2.1 Supervised Learning

As was already said, in supervised learning, a trainer provides input and output examples, and the student is responsible for independently adjusting the system's parameters so that it would produce the right output pattern when stimulated by one of the provided input patterns [15–17]. In this part, we'll discuss two key supervised learning styles: inductive learning and analogous learning [18–21]. In the following chapter, a number of other supervised learning methods utilizing neural nets will be discussed. Figure 6.2 shows the different types of supervised learning methods.

#### *6.2.1.1 Inductive Learning*

When looking at inductive learning, the challenge is to predict the function ($f$) for given input samples ($x$) and output samples ($f(x)$). The challenge is to extrapolate from the samples and mapping in a way that will be beneficial for estimating the output for fresh samples in the future [22, 23]. A hypothesis can be developed $h(x_i) \gg f(x_i)$ for a given a set of $x_i$, and $f(x_i)$ pairs using the supervised learning techniques known as inductive learning.

According to the nature of the $\{x_i, f(x_i)\}$ dataset, we may employ the neural learning techniques. The learning algorithm for such numerical sets $\{x_i, f(x_i)\}$ must be able to adapt the parameters of the learner much effective than the curve fitting techniques. The amount of adaptations will increase with the number of training instances. Both learning by decision tree (DT) and learning by version space are significant subtypes of inductive learning.

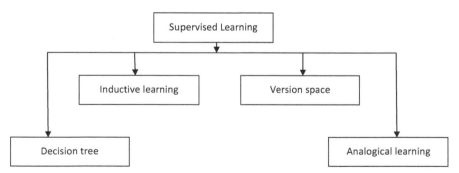

**FIGURE 6.2**
Types of supervised learning algorithm.

### 6.2.1.2 Learning by Version Space

Vector space learning technique is one of the earliest types of inductive learning (IL) proposed by Mitchell in the early 1980s [24, 25]. All the helpful information can be recall by hierarchical representation of knowledge of a version space. The version space approach involves managing many models within a version space to facilitate concept learning [26, 27].

**Characteristics**

- Version spaces are used to convey tentative heuristics.
- A version space is a representation of every possible heuristic description.
- A description is considered credible if it holds true for all known positive examples while holding true for no known negative cases.
- Two complementing trees make up a version space description: one with nodes related to overly general models, the other with nodes connected to highly specialized models.

### 6.2.1.3 Learning by Decision Tree (DT)

A decision tree produces a binary judgment of true or false values after receiving a set of attributes (or properties) of the objects as inputs [28]. As a result, decision trees typically represent Boolean functions. Continuous values of the output parameters are permitted in addition to a 0–1 range [29]. However, we assume the limitation to Boolean outputs for the sake of simplicity. A decision tree's nodes each represent a test of an instance's attribute, and each branch descending from that node represents one of the attribute's potential values [28, 30]. We take into account a number of cases, some of which produce a true value for the choice, to demonstrate the contribution of a decision tree. The previous one is the positive example. On the other side, we refer to an instance as "a negative instance" when the outcome is a wrong decision [31]. Now let's look at the issue of a bird learning to fly. Figure 6.3 shows the diagram for whether a bird will be able to fly or not.

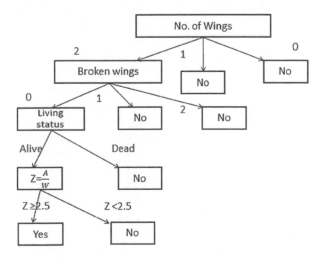

**FIGURE 6.3**
Decision tree for a bird learning to fly or not.

### 6.2.1.4 Analogical Learning

A problem can have both positive and negative examples, and in inductive learning, the learner must develop a notion that encompasses the majority of the positive examples while excluding any negative ones [32, 33]. This indicates that inductive learning requires a number of training experiences in order to establish an idea. Analogical learning, however, just requires one example to be successful. For instance, one must ascertain the plural form of "bacillus" given the following training example [34, 35]. By substituting "i" for "us," the analogical learning system learns that words ending in "us" take on the plural form.

Here are the formalized main steps of analogy learning.

1. Identifying analogy: Determine whether a new problem and an experienced problem instance are comparable.
2. Establishing the mapping function: The mapping is established after relevant components of the encountered problem are chosen.
3. Implementation of mapping function: A specific domain can be switched to a new domain using the mapping function.
4. Validation: Through trial procedures like theorem or simulation, the newly developed solution is validated for its applicability [34].
5. Learning: After the validation is over, new data is encoded and saved for further use.

### 6.2.2 Unsupervised Learning

In a learning algorithm, when input and output problems are provided, the learner is required to build a mapping function that yields the right output for a specific input pattern [36, 37]. But there is no trainer involved in unsupervised learning. So, the learner needs to implement the methodology by experimenting on the environment. The environment is responsive, but it does not distinguish between activities that are rewarding and those that are punitive. Because the objectives or results of the training information are unclear, the environment is unable to assess how the learner's activities are progressing toward the objectives [38]. Experiments are one of the simplest ways to build concepts through unsupervised learning. Consider the following scenario: A toddler throws a ball at a wall; the ball bounces and comes back to them. The toddler learns the "principle of bouncing" after doing this experiment several times. Of course, this is an illustration of unsupervised learning. The majority of scientific laws were created by this algorithm [38–40]. To demonstrate the fundamentals of concept development through unsupervised learning, let's look at another example. Consider that we want to group animals according to their height-to-weight proportion to the speed. For instance, we measure the aforementioned characteristics of sample animals and plot them on a two-dimensional frame. Figure 6.4 shows that tigers, foxes, cows, and dogs belong to various classes. Additionally, there is overlap between the dog and fox classifications. Now, if we are given a measured number for an unknown animal's speed and height/weight ratio, we can readily identify it—as long as it does not fall into an overlapped category [39]. Because it lacks the necessary traits to accurately characterize the animal, an overlapped region cannot identify the creatures. For instance, the speed and height/weight ratio of foxes and dogs are comparable. Other characteristics, such as face shape, are also needed to distinguish between them.

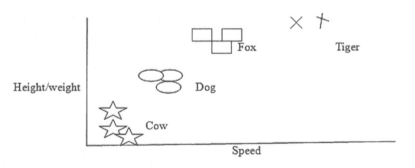

**FIGURE 6.4**
Learning animals through classification.

In many cases, it might be difficult to identify the features themselves, especially when the problem's characteristics are unclear. We take one of the biological classification issues as an example.

The classification of patterns is crucial in practically every field of science, not only biology. For instance, in psychology, pattern classification is used to categorize patients with various mental illnesses so that they can be treated with a single therapy [39]. Prior to matching a suspect's fingerprint with databases of that class that are already known, criminologists classify fingerprints into common categories. To extract the features from the dataset is the first stage in pattern classification [65].

### 6.2.3 Reinforcement Learning

In reinforcement learning, the learner adjusts its parameters based on the environment's feedback signal status (reward or punishment) [41, 42]. Learning automata use the simplest type of reinforcement learning. The feedback signal exhibited in Figure 6.5 shows the reward/punishment status has led to the development of Q-learning and temporal difference learning [43].

#### 6.2.3.1 Learning Automata

The most popular of the well-known reinforcement learning techniques is the learning automaton [44]. Two modules make up the learning mechanism of such a system: the environment and the learning automation. The creation of a stimulus by the environment initiates the learning cycle. The automation reacts to the surroundings after getting

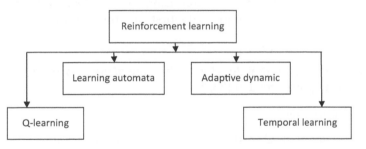

**FIGURE 6.5**
Types of reinforcement learning.

a stimulus [45]. The environment takes in the response, assesses it, and provides the automation with a fresh stimulus. Based on the most recent answer and the current input (stimulus) of the automation, the learner then automatically modifies its parameters [46]. Numerous applications to challenges in the actual world can be made using the learning automata's guiding principles.

### 6.2.3.2 Adaptive Dynamic Programming

The agent is assumed to have received a response from the environment in the case of reinforcement learning, but the agent can only identify its status—whether it is rewarding or punishing—at the terminal stage of its activity [47, 48]. We also assume that the agent starts off in state $S_0$ and changes to state $S_1$ after acting on the environment. The agent changes its state from $S_0$ to $S_1$ as a result of action $a_0$, which is indicated by the symbol $a_0$. A utility function can also be used to represent an agent's reward. A Ping-Pong agent's utility, for instance, might be one of its selling factors [49]. In reinforcement learning, the agent could be passive or active. A passive learner makes an effort to understand the utility by considering it in various stages. On the other hand, a student that is an active learner can extrapolate the usefulness at unknown stages from the knowledge it has acquired.

### 6.2.3.3 Q-learning

One of the off-policy strategies and most widely used reinforcement learning approaches is Q-learning [50]. Early Q-learning algorithms only supported a limited set of applications and were unsatisfactory in a number of ways. It has also been noted that this powerful algorithm occasionally learns unreasonably and overestimates the values of the actions taken, which lowers overall performance [51]. Recent developments in machine learning have led to the discovery and widespread use of more Q-learning variations, such as deep Q-learning, which blends fundamental Q-learning with deep neural networks (DNNs) [52].

#### 6.2.3.3.1 Basic Q-Learning

This type of Q-learning algorithm distinguishes the acting policy from the learning policy, in contrast to previous that unable to distinguish between behavior and learning [51].

#### 6.2.3.3.2 Deep Q-Learning

Convolution neural networks (CNNs) and fundamental Q-learning are combined to create deep Q-learning, which was created by Google Deep Mind. When expressing the value function for each state becomes challenging, QL uses a CNN to approximate the function [51]. In addition to the value approximation using a CNN, deep Q-learning includes two techniques [52]. The target Q technique is the first, and the second is an experience replay.

#### 6.2.3.3.3 Hierarchical Q-Learning

In order to address the issues that develop as the state-action space of Q-learning expands, hierarchical Q-learning was created [53, 54].

#### 6.2.3.3.4 Double Q-Learning

To address the issue that Q-learning struggles in a stochastic environment, double Q-learning was created. Q-learning is biased in a stochastic setting [55] because the agent's action value is overstated.

*6.2.3.3.5 Multi-Agent*

In order to address the issue of basic Q-learning inefficiency in multi-agent systems, multi-agent modular Q-learning was developed [56]. According to [57] the state space for each agent grows exponentially in size as the number of agents increases. The quantity of memory and states could explode.

### 6.2.3.4 Temporal Difference Learning

In order to forecast the total reward anticipated over the long term, temporal difference learning (TD), an unsupervised learning technique, is frequently employed in reinforcement learning (RFL) [58, 59]. However, it can also be used to forecast the test dataset, unlike other algorithms. In essence, this method identifies the potential input parameter before forecasting a new dataset. Using a succession of intermediate incentives, the long-term value of a behavior pattern can be calculated using the TD learning method. Future predictions serve as the training signal for predictions in TD learning. This approach combines the Monte Carlo (MC) and dynamic programming (DP) approaches [60–62]. Temporal difference approaches typically alter forecasts to match subsequent, more accurate predictions for the future, much before the ultimate conclusion is evident and known, in contrast to Monte Carlo methods, which adjust their estimates only after the final result is known.

### 6.2.4 Learning by Inductive Logic Programming (ILP)

The branch of machine learning known as inductive logic programming (ILP) uses first-order logic (FOL) to express hypotheses and data. ILP primarily addresses issues with structured data and background knowledge since FOL is expressive and declarative [63]. With the help of "upgrades" of current propositional machine learning systems, inductive logic programming addresses a wide range of machine learning issues, such as clustering, classification, reinforcement learning, and regression. For the purposes of knowledge representation and reasoning, it uses logic. Applications like web mining, natural language processing, and bio- and chemo-informatics have all benefited from the use of ILP systems. ILP's primary and most significant benefit is that it gets beyond attribute-value learning systems' representational constraints [64]. The use of ILP is also encouraged by the fact that it makes use of the declarative logic language. It is suggested that theories are interpretable and comprehensible.

## 6.3 Summary

The four main categories of machine learning algorithms are covered in this chapter, along with some of their subcategories. Learning automata, decision trees, inductive logic programming, and version space have all been used to introduce the ideas of supervised learning. Unsupervised machine learning concepts are explained with an appropriate example. Reinforcement learning, which is more recent than supervised machine learning and has been extensively studied by Q-learning and its subcategories as well as temporal difference learning, is one of the most recent parts of learning. Machine learning research is still being done in the domain of inductive logic programming (ILP). The theory may be expanded in the future to uncover information in real-world systems.

# References

1. Zhao Y, Smidts C. CMS-BN: A cognitive modeling and simulation environment for human performance assessment, Part 1—Methodology. *Reliability Engineering & System Safety* 2021;213: 107776.

2. Amit K. *Artificial intelligence and soft computing: Behavioral and cognitive modeling of the human brain*. CRC Press;2018.

3. Lee MD. Bayesian methods in cognitive modeling. *The Stevens' Handbook of Experimental Psychology and Cognitive Neuroscience* 2018;5: 37–84.

4. Lee MD, Criss AH, Devezer B, Donkin C, Etz A, Leite FP, ... Vandekerckhove J. Robust modeling in cognitive science. *Computational Brain & Behavior* 2019;2(3): 141–153.

5. Battineni G, Sagaro GG, Chinatalapudi N, Amenta F. Applications of machine learning predictive models in the chronic disease diagnosis. *Journal of Personalized Medicine* 2020;10(2): 21.

6. Zhu X, Singla A, Zilles S, Rafferty AN. An overview of machine teaching. 2018. arXiv:1801.05927.

7. Fisher CK, Smith AM, Walsh JR. Machine learning for comprehensive forecasting of Alzheimer's disease progression. *Scientific Reports* 2019;9(1): 1–14.

8. Conati C, Porayska-Pomsta K, Mavrikis M. AI in education needs interpretable machine learning: Lessons from open learner modelling. 2018. arXiv:1807.00154.

9. Mohammadi M, Al-Fuqaha A. Enabling cognitive smart cities using big data and machine learning: Approaches and challenges. *IEEE Communications Magazine* 2018;56(2): 94–101.

10. Goh YM, Ubeynarayana CU, Wong KLX, Guo BH. Factors influencing unsafe behaviors: A supervised learning approach. *Accident Analysis & Prevention* 2018;118: 77–85.

11. Bourgin DD, Peterson JC, Reichman D, Russell SJ, Griffiths TL. Cognitive model priors for predicting human decisions. *International Conference on Machine Learning*. PMLR;2019, May, pp. 5133–5141.

12. Lv Z, Qiao L, Singh AK. Advanced machine learning on cognitive computing for human behavior analysis. *IEEE Transactions on Computational Social Systems* 2020;8(5): 1194–1202.

13. Tremblay C, Aladin S. Machine learning techniques for estimating the quality of transmission of lightpaths. *Proceedings of the 2018 IEEE Photonics Society Summer Topical Meeting Series (SUM)*. IEEE;2018, July, pp. 237–238.

14. Nakamura T, Nagai T, Taniguchi T. Serket: An architecture for connecting stochastic models to realize a large-scale cognitive model. *Frontiers in Neurorobotics* 2018;12: 25.

15. Dutta P, Paul S, Kumar A. Comparative analysis of various supervised machine learning techniques for diagnosis of COVID-19. In *Electronic devices, circuits, and systems for biomedical applications*, Hristu-Varsakelis D, Levine WS (Eds.). Academic Press;2021, pp. 521–540.

16. Dutta P, Paul S, Obaid AJ, Pal S, Mukhopadhyay K. Feature selection based artificial intelligence techniques for the prediction of COVID like Diseases. *Journal of Physics: Conference Series* 2021, July;1963(1): 012167.

17. Dutta P, Paul S, Shaw N, Sen S, Majumder M. Heart disease prediction: A comparative study based on a machine-learning approach. In *Artificial intelligence and cybersecurity*. CRC Press;2022, pp. 1–18.

18. Dutta P, Paul S, Majumder M. *Intelligent smote based machine learning classification for fetal state on cardiotocography dataset*. 2021.

19. Dutta P, Paul S, Majumder M. *An efficient smote based machine learning classification for prediction & detection of PCOS*. 2021.

20. Sen S, Saha S, Chaki S, Saha P, Dutta P. Analysis of PCA based AdaBoost machine learning model for predict mid-term weather forecasting. *Computational Intelligence and Machine Learning* 2021;2(2): 41–52.

21. Duttaa P, Shawb N, Dasc K, Ghosh L. *Early & accurate forecasting of mid term wind energy based on PCA empowered supervised regression model* 2021;2(2): 53–64.

22. Zeng H, Zhou H, Srivastava A, Kannan R, Prasanna V. Graphsaint: Graph sampling based inductive learning method. 2019. arXiv:1907.04931.

23. Madala HR *Inductive learning algorithms for complex systems modeling*. CRC Press;2019.
24. Rane S, Brito AE. A version space perspective on differentially private pool-based active learning. *Proceedings of the 2019 IEEE International Workshop on Information Forensics and Security (WIFS)*. IEEE;2019, December, pp. 1–6.
25. Chen Y, Singla A, Mac Aodha O, Perona P, Yue Y. Understanding the role of adaptivity in machine teaching: The case of version space learners. *Advances in Neural Information Processing Systems* 2018;31: 1–11.
26. Palma L, Diao Y, Liu A. *Efficient version space algorithms for "human-in-the-loop"*. *Model Development*;2020.
27. Groza A, Ungur I. Improving conflict resolution in version spaces for precision agriculture *International Journal of Agricultural Science* 2018;3.
28. Charbuty B, Abdulazeez A. Classification based on decision tree algorithm for machine learning. *Journal of Applied Science and Technology Trends* 2021;2(01): 20–28.
29. Patel HH, Prajapati P. Study and analysis of decision tree based classification algorithms. *International Journal of Computer Sciences and Engineering* 2018;6(10): 74–78.
30. Grąbczewski K. *Meta-learning in decision tree induction*. Springer International Publishing;2014, Vol. 1.
31. Kamwa I, Samantaray SR, Joos G. Catastrophe predictors from ensemble decision-tree learning of wide-area severity indices. *IEEE Transactions on Smart Grid* 2010;1(2): 144–158.
32. Gentner D. Analogical learning. *Similarity and Analogical Reasoning* 1989;199: 1–66.
33. Chen K, Rabkina I, McLure MD, Forbus KD. Human-like sketch object recognition via analogical learning. *Proceedings of the AAAI Conference on Artificial Intelligence*. 2019, July, Vol. 33, pp. 1336–1343.
34. Gentner D, Smith LA. Analogical learning and reasoning. In *The Oxford handbook of cognitive psychology*, Reisberg D (Ed.). Oxford University Press;2013, pp. 668–681.
35. Brown AL. Analogical learning and transfer: What develops. In *Similarity and analogical reasoning*, Vosniadou S, Ortony A (Eds.). Cambridge University Press;1989, pp. 369–412.
36. Li N, Shepperd M, Guo Y. A systematic review of unsupervised learning techniques for software defect prediction. *Information and Software Technology* 2020;122: 106287.
37. Pinheiro O, Almahairi PO, Benmalek A, Golemo RF, Courville AC. Unsupervised learning of dense visual representations. *Advances in Neural Information Processing Systems* 2020;33: 4489–4500.
38. Caron M, Misra I, Mairal J, Goyal P, Bojanowski P, Joulin A. Unsupervised learning of visual features by contrasting cluster assignments. *Advances in Neural Information Processing Systems* 2020;33: 9912–9924.
39. Berry MW, Mohamed A, Yap BW (Eds.). *Supervised and unsupervised learning for data science*. Springer Nature;2019.
40. Caron M, Bojanowski P, Joulin A, Douze M. Deep clustering for unsupervised learning of visual features. *Proceedings of the European Conference on Computer Vision (ECCV)*. 2018, pp. 132–149.
41. Levine S, Kumar A, Tucker G, Fu J. Offline reinforcement learning: Tutorial, review, and perspectives on open problems. 2020. arXiv:2005.01643.
42. François-Lavet V, Henderson P, Islam R, Bellemare MG, Pineau J. An introduction to deep reinforcement learning. *Foundations and Trends® Machine Learning* 2018;11(3–4): 219–354.
43. Botvinick M, Ritter S, Wang JX, Kurth-Nelson Z, Blundell C, Hassabis D. Reinforcement learning, fast and slow. *Trends in Cognitive Sciences* 2019;23(5): 408–422.
44. Zhang Z, Wang D, Gao J. Learning automata-based multiagent reinforcement learning for optimization of cooperative tasks. *IEEE Transactions on Neural Networks and Learning Systems* 2020;32(10): 4639–4652.
45. Ayanzadeh R, Halem M, Finin T. Reinforcement quantum annealing: A hybrid quantum learning automata. *Scientific Reports* 2020;10(1): 1–11.
46. Rezvanian A, Moradabadi B, Ghavipour M, Khomami MMD, Meybodi MR. *Learning automata approach for social networks*. springer;2019, Vol. 820.
47. Liu D, Xue S, Zhao B, Luo B, Wei Q. Adaptive dynamic programming for control: A survey and recent advances. *IEEE Transactions on Systems, Man, and Cybernetics: Systems* 2020;51(1): 142–160.

48. Yang Y, Vamvoudakis KG, Modares H, Yin Y, Wunsch DC. Hamiltonian-driven hybrid adaptive dynamic programming. *IEEE Transactions on Systems, Man, and Cybernetics: Systems* 2020;51(10): 6423–6434.

49. Lu J, Wei Q, Wang FY. Parallel control for optimal tracking via adaptive dynamic programming. *IEEE/CAA Journal of Automatica Sinica* 2020;7(6): 1662–1674.

50. Kumar A, Zhou A, Tucker G, Levine S. Conservative Q-learning for offline reinforcement learning. *Advances in Neural Information Processing Systems* 2020;33: 1179–1191.

51. Fan J, Wang Z, Xie Y, Yang Z. A theoretical analysis of deep Q-learning. In *Learning for dynamics and control*, A. Bayen, A. Jadbabaie, G. J. Pappas, P. Parrilo, B. Recht, C. Tomlin, M. Zeilinger (Eds.). PMLR;2020, July, pp. 486–489.

52. Jang B, Kim M, Harerimana G, Kim JW. Q-learning algorithms: A comprehensive classification and applications. *IEEE Access* 2019;7: 133653–133667.

53. Xu B, Zhou Q, Shi J, Li S. Hierarchical Q-learning network for online simultaneous optimization of energy efficiency and battery life of the battery/ultracapacitor electric vehicle. *Journal of Energy Storage* 2022;46: 103925.

54. Liu J, Sha N, Yang W, Tu J, Yang L. Hierarchical Q-learning based UAV secure communication against multiple UAV adaptive eavesdroppers. *Wireless Communications and Mobile Computing* 2020;2020: 1–15.

55. Xiong H, Zhao L, Liang Y, Zhang W. Finite-time analysis for double Q-learning. *Advances in Neural Information Processing Systems* 2020;33: 16628–16638.

56. Wang J, Ren Z, Liu T, Yu Y, Zhang C. Qplex: Duplex dueling multi-agent Q-learning. 2020. arXiv preprint arXiv:2008.01062.

57. Matta M, Cardarilli GC, Di Nunzio L, Fazzolari R, Giardino D, Re M, Spanò S. Q-RTS: A real-time swarm intelligence based on multi-agent Q-learning. *Electronics Letters* 2019;55(10): 589–591.

58. Bhandari J, Russo D, Singal R. A finite time analysis of temporal difference learning with linear function approximation. In *Conference on learning theory*. PMLR;2018, July, pp. 1691–1692.

59. Cai Q, Yang Z, Lee JD, Wang Z. Neural temporal-difference learning converges to global optima. *Advances in Neural Information Processing Systems* 2019;32: 1–12.

60. Ghiassian S, Patterson A, Garg S, Gupta D, White A, White M. Gradient temporal-difference learning with regularized corrections. In *International Conference on Machine Learning*. PMLR;2020, November, pp. 3524–3534.

61. Janner M, Mordatch I, Levine S. Gamma-models: Generative temporal difference learning for infinite-horizon prediction. *Advances in Neural Information Processing Systems* 2020;33: 1724–1735.

62. Lee D, He N. Target-based temporal-difference learning. In *International Conference on Machine Learning*. PMLR;2019, May, pp. 3713–3722.

63. Cropper A, Dumančić S. 2020. arXiv: 2008.07912.

64. Cropper A, Dumančić S, Evans R, Muggleton SH. Inductive logic programming at 30. *Machine Learning* 2022;111(1): 147–172.

65. Liu Y, Zhang D, Lu G. Region-based image retrieval with high-level semantics using decision tree learning. *Pattern Recognition* 2008;41(8): 2554–2570.

# Part B

# Artificial Intelligence and Cognitive Computing

*Practices*

# 7

# Parametric Optimization of n-Channel JFET Using Bio Inspired Optimization Techniques

## 7.1 Introduction

The most foundational and prevalent device is the field effect transistor in electronic circuit applications [1, 2]. In this type of transistor electric field control follows the current passing through the three terminal semiconductors [3]. For an n-channel junction field effect transistor (JFET) the nature of substrate is n-type semiconductor. Current–voltage (I-V) characteristics of JFET can be defined by drain characteristics (where $V_{DS}$ regulates the current flow $I_D$) and transfer characteristics ($V_{GS}$ regulates the current flow $I_D$). In drain characteristics, the current of a JFET typically depends on supply voltage on drain terminal ($V_{DD}$) and drain resistance (Rd), and in transfer characteristics, drain current depends upon the $I_{DSS}$ and pinch-off voltage ($V_P$) [4, 5]. In each of the cases one set of control parameters is kept constant during the optimization of model parameters. In order to improve, inspect, and assess the effectiveness of JFET-based systems, engineers must have a precise understanding of the n-channel JFET transistor characteristics from experimental evidence. This requires creating a high precision quantitative model to represent the nonlinear I-V correlation of the FET. Modeling approach for JFETs is essentially an optimization issue that is nonlinear [6].

Over the past few years, ensemble methods—usually motivated by objective truths, animal behavior, or developmental concepts—have grown highly popular [7, 8]. Swarm intelligence, however, is the most common and effective class of metaheuristic algorithms (SI). The genetic algorithm (GA) [9, 10], particle swarm optimization (PSO) [11], cuckoo search optimization (CSO) [12], gray wolf optimization (GWO) [13], ant colony optimization (ACO) [14, 15], bat algorithm (BA) [16, 17], elephant swarm water search algorithm (ESWSA) [18, 19], firefly algorithm (FA) [20, 21], artificial bee colony optimization (ABCO) [22, 23], flower pollination algorithm (FPA) [24–26], and differential evolution (DE) [11, 27], among others, are examples of such kinds of metaheuristic. In literature, many researchers used several metaheuristics like BA, CSO, FA, FPA, evolutionary algorithm, GWO and ABCO for the parameter estimation problems of different semiconductor devices. No one has attempted to use an optimization algorithm to identify the parameters for n-channel JFET drain and transfer character traits. One newly developed, effective, and well-liked SI-based metaheuristic approach that draws inspiration from flower pollination is the FPA. FPA has so far been effectively used to solve a variety of global optimization, multimodal optimization, limited optimization, structural engineering, and reverse engineering problems, among other problems.

DOI: 10.1201/9781003216001-9

In this research, three bio inspired optimization techniques—FPA, PSO, and CSO—are used for parametric optimization of drain and transfer characteristics of n-channel JFET. The remainder of the chapter is arranged as follows. The mathematical modeling of the parameter evaluation issue and the FPA approach are described in Section 7.2. The discussion moves on to experimental data, methods, and simulation outcomes. References are listed after the conclusion.

## 7.2 Mathematical Description

### 7.2.1 Current Equation for JFET

The current ($I$) in a drain characteristic of an n-channel JFET [28] can be given as in Eq. (7.1):

$$V_{DS} = V_{DD} - I_D R_D$$

$$V_{DS} = V_{DD} - R_D * I_{DSS} \left( 1 - \frac{V_{GS}}{V_P} \right)^2 \tag{7.1}$$

Current ($I$) in a transfer characteristic of an n-channel JFET can be given as in Eq. (7.2):

$$I_D = I_{DSS} \left( 1 - \frac{V_{GS}}{V_P} \right)^2 \tag{7.2}$$

Where, $I_D$ is drain current, $I_{DSS}$ is the saturated drain current, $V_{DD}$ is supply voltage at the drain terminal, $V_{DS}$ is drain to source voltage, $V_P$ is pinch-off voltage, and $R_D$ is drain resistance. So, this drain characteristics of JFET model contains two parameters $\left( V_{DD} \text{ and } R_D \right)$, and the transfer characteristics model contains two parameters $\left( I_{DSS}, V_P \right)$ to be estimated.

### 7.2.2 Flower Pollination Algorithm

According to Abdel-Basset and Shawky [29] and Nguyen et al. [30], flower pollination is generally related to the transportation of pollen enabling vegetative propagation, with insects, birds, and bats serving as the primary pollinators for this transference. A recently suggested metaheuristic called FPA [31] uses certain parts as criteria for pollination. Pollinators that convey pollen replicate *Lévy* flights during transit, which is consistent with the assumption that biological cross-pollination is a mechanism of global pollination (Rule 1). Abiotic pollination and self-pollination are exploited for local pollination (Rule 2). Depending on how closely two blooms resemble one another, or the likelihood of successful reproduction, pollinators may acquire floral dependability (Rule 3). A switch probability p ε [0, 1] that is slightly skewed in favor of local pollination can be used to regulate the transition from local to global pollination (Rule 4). Each bloom or pollen in this case represents a remedy to an optimization issue. The two following equations represent global and local pollination, or search, respectively:

$$x_i^{t+1} = x_i^t + \gamma L\acute{e}vy\lambda \left( g * - x_i^t \right) \tag{7.3}$$

$$x_i^{t+1} = x_i^t + \varepsilon\left(x_j^t - x_k^t\right) \tag{7.4}$$

Where it is the ultimate idea currently identified among all possibilities at the latest incarnation, $x_j^t$ and $x_k^t$ are pollens from separate flowers of the same vegetation types, $\varepsilon$ stands for random walk step size, and $\gamma$ is a scaling factor stands for randomness step size $g*$ within a homogenous distribution in [0,1]. $x_i^t$ is the pollen $i$ or solution vector $x_i$ at current iteration. FPA was chosen as an optimization approach because it provides higher resolution and fidelity than other well-known metaheuristic methods [28, 32].

### 7.2.3 Objective Function

To assess the quality of a solution, all optimal control techniques employ a performance index or fitness value. In order to reduce the discrepancy and simulated currents, the estimate job seeks the most ideal values for the random variables. The root mean square of the error (RMSE) is defined as Equations (7.5) and (7.6) can be used as the objective function [13].

$$\text{RMSE}(X) = \sqrt{\frac{1}{N}\sum_{i=1}^{n} f(V_{\text{DD}}, R_D, X)} \tag{7.5}$$

for drain characteristics

$$\text{RMSE}(X) = \sqrt{\frac{1}{N}\sum_{i=1}^{n} f(I_{\text{DSS}}, V_p, X)} \tag{7.6}$$

for transfer characteristics
Where N is the quantity of empirical observations, that is, a set of n-channel JFET voltage and current, $X$ is the set of the estimated parameters, that is, $X = V_{\text{DD}}, R_D$ for drain characteristics and $X = I_{\text{DSS}}, V_p$ for transfer characteristics. Now three bio-inspired optimization techniques, PSO, FPA, and CSO, are employed to reduce the value of the aforementioned function, so that the best value of $X = V_{\text{DD}}, R_D$ for drain characteristics and $X = I_{\text{DSS}}, V_p$ for transfer characteristics can be obtained.

## 7.3 Methodology

The two following phases make up the primary portions of the overall process of metaheuristic-based optimization of n-channel JFET model parameters: (1) Research in the lab led to the observation of a number of JFET voltages and matching drain currents in drain and transfer characteristics. (2) Implementation of PSO, FPA, and CSO to optimize the model parameters of JFET. Table 7.1 represents the IC specification of N-channel JFET. The details of these steps are explained in Figure 7.2.

For the laboratory experiment, J112A n-channel JFET and circuit connection shown in Figure 7.1. Then we apply two variable DC power supplies across them in the input circuit

**TABLE 7.1**

J112A N-Channel JFET Specification

| Gate–Source Breakdown Voltage $V_{GS}$ (V) | Gate-Source Cut Off Voltage $V_{GS-off}$ (V) | Zero Gate Voltage Drain Current $I_{DSS}$ (mA) | Drain-Source Resistance $R_{DS}$ (ohm) | Operating Temperature Range (°C) | Package |
|---|---|---|---|---|---|
| 35 | −5 | 5 | 50 | −65–150 | Metal TO-92 |

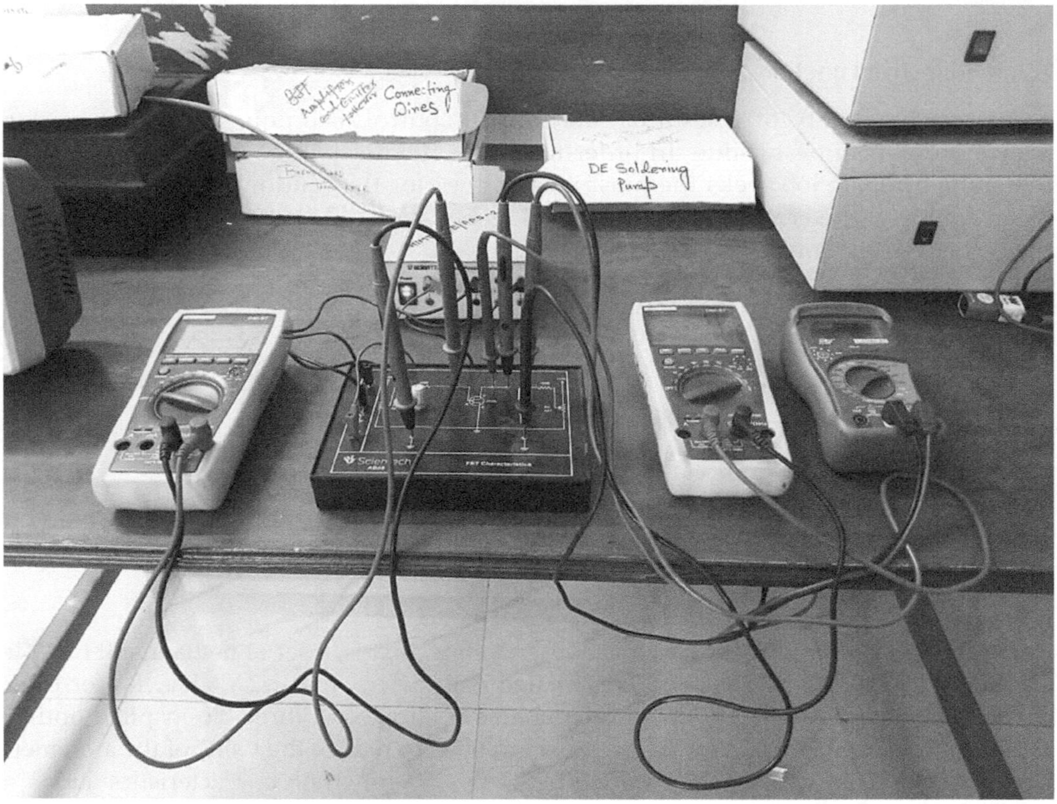

**FIGURE 7.1**
Circuit diagram for determining the drain and transfer characteristics of n-channel JFET.

for reverse bias $V_{GS}$ and the output circuit for forward bias operation $V_{DS}$. Next, (1) in the case of drain characteristics, by keeping $V_{GS}$ constant, the output circuit voltage $V_{DS}$ is progressively raised and a commensurate drain current is shown as a result; and (2) in transfer characteristics, by keeping output control voltage $V_{DS}$ constant, $V_{GS}$ is gradually increased and the corresponding drain is measured.

In next phase of this work, FPA has been used for optimization of n-channel JFET parameters. In our present problem, the dimension of search for the metaheuristic is 2 as the input variables of optimization process are $X = V_{DD}, R_D$ for drain characteristics and $X = I_{DSS}, V_p$ for transfer characteristics. Instruction and the computation of the genetic algorithm are done using observational evidence. RMSE is employed as the fitness function. The population and overall number of iterations for PSO, FPA, and CSO are set as 100 and 1000, respectively. The lower and upper limits, that is, the search range, for supply voltage

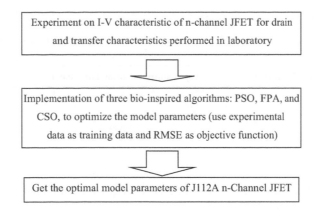

**FIGURE 7.2**
Stepwise methodologies for the present research.

and drain resistance are chosen as [6 V, 12 V] and [1 Kohm, 10 Kohm] for drain characteristics, and the zero gate voltage drain current and pinch-off voltage are chosen as [10 mA, 15 mA] and [3 V, 5 V], respectively, for transfer characteristics. Following iteration, the best optimization strategy will arrive at the optimal solution, or the best set of process variables that minimizes RMSE. Calculated I-V characteristics should be almost identical to the diode's experimental I-V characteristics under ideal conditions.

## 7.4 Result and Discussion

The findings of this work have been demonstrated and discussed in this part in order to make some significant conclusions. Initially, a total of 15 readings of drain characteristics for three different sets of control voltage $V_{GS}$ (0 V, –1 V, and –1.5 V), voltage $V_{DS}$, and current $I_D$ have been observed, while in transfer characteristics a total of 17 readings were taken for $V_{GS}$ and current $I_D$ in n-channel JFET. The experimental I-V characteristic of a p-n junction diode is shown in Figure 7.3 and is almost exponential in nature. Next, three bio-inspired algorithms, PSO, FPA, and CSO, have been applied to this experimental dataset to find the optimal values of supply voltage and drain resistance and the zero gate voltage drain

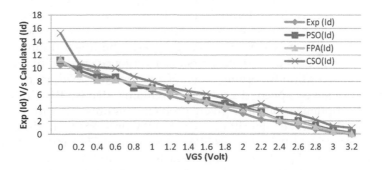

**FIGURE 7.3**
Transfer characteristics of n-channel JFET.

**TABLE 7.2**

Comparative Study Based Accuracy and Average Computational Time in Drain Characteristics

| Name of the Algorithm | Avg. Computational Time (sec) | RMSE (%) |
|---|---|---|
| PSO | 256.21 | 5.42 |
| FPA | 129.32 | 3.41 |
| CSO | 343.652 | 9.18 |

**TABLE 7.3**

Comparative Study Based Accuracy and Average Computational Time in Transfer Characteristics

| Name of the Algorithm | Avg. Computational Time (sec) | RMSE (%) |
|---|---|---|
| PSO | 221.41 | 4.32 |
| FPA | 147.31 | 2.19 |
| CSO | 289.54 | 7.52 |

current and pinch-off voltage for transfer characteristics. All of the bio-inspired algorithms have been run 10 times using the configuration described previously since they provide various results based on startup and search method unpredictability. The final results were then obtained by a statistical analysis. The fitness value, or RMSE [33, 34], is shown in Tables 7.2 and 7.3 for each program run. It is evident that FPA is always capable of achieving the minimum values of 0.00341 and 0.00219 for drain and transfer characteristics or those close by. The overall performance of FPA for this JFET optimization task can thus be seen as being quite excellent.

Computational time [35] indicates the time taken by the optimization algorithm to minimum fitness value of the objective function. Table 7.2 and Table 7.3 show the computational time taken by the applied optimization algorithms. In both cases FPA took the least computational time to provide the least fitness value of drain current fitness function (Figures 7.4 and 7.5).

The predicted value of the drain current was then determined using Equations (7.1) and (7.2) along with the aforementioned parameters. Figure 7.3's comparison of the observed I-V characteristics of n-channel JFETs with computed output as shown in Figure 7.6. Figure 7.6 shows that the I-V attributes of an n-channel JFET between experimental evidence and computed or simulated values change very little. This claim certifies a technique that is suggested for determining the JFET's ideal parameters, with the goal of minimizing the difference between estimated and empirical values of diode current. From the characteristics graphs it is seen that FPA generates the fittest drain current over the other algorithms while CSO performed worst. Convergence speed [36] is another feature to identify the best algorithm for optimization of any function. With increasing number of iterations, if convergence speed decreases or remains steady, this indicates the best algorithm for the optimization of a given problem. Figures 7.7 and 7.8 represent the convergence speed of three algorithms, CSO, FPA and PSO, in drain and transfer characteristics, respectively. In both cases FPA outperformed CSO and PSO. In this research we consider the set of $X = V_{DD}, R_D$ for drain characteristics and $X = I_{DSS}, V_p$ for transfer characteristics corresponding to the best fitness, that is, least RMSE, as the final output. Table 7.4 shows the optimal parameter values of J112A n-Channel JFET at the two above-mentioned conditions.

**TABLE 7.4**

Optimal Values for Drain and Transfer Characteristics at Different Conditions

| Algorithm | Transfer Characteristics | | Drain Characteristics | |
|---|---|---|---|---|
| | $I_{DSS}$ (mA) | $V_p$ (volt) | $V_{DD}$ (volt) | $R_D$ (k) |
| CSO | 10.98 | 4.9 | 11.9 | 4.8 |
| FPA | 12 | 5 | 12 | 5 |
| PSO | 11.85 | 5 | 12 | 5 |

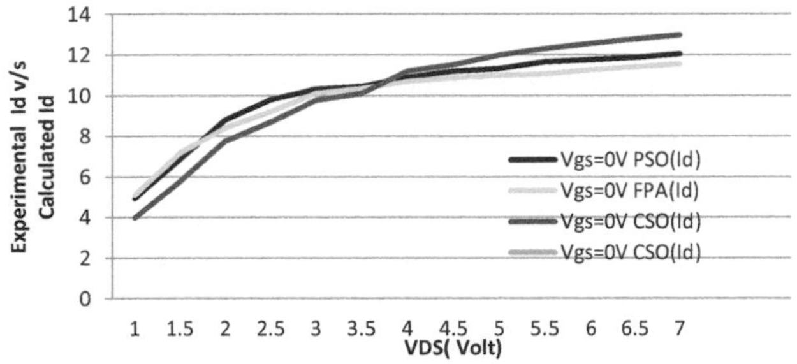

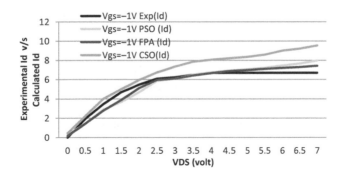

**FIGURE 7.4**
Drain characteristics when VGS is set at 0 V.

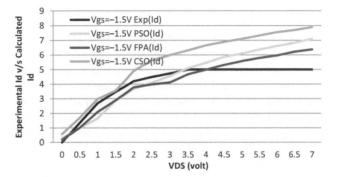

**FIGURE 7.5**
Drain characteristics when VGS is set at −1 V.

**FIGURE 7.6**
Drain characteristics when VGS is set at −1.5 V.

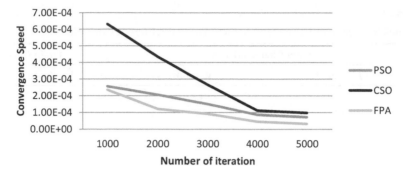

**FIGURE 7.7**
Convergence speed of the proposed algorithm for drain characteristics.

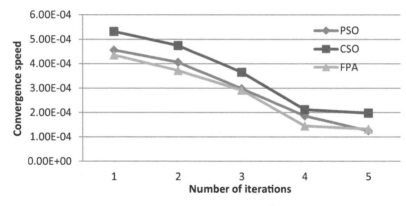

**FIGURE 7.8**
Convergence speed of the proposed algorithm for transfer characteristics.

## 7.5 Conclusion

An essential challenge in the field of electronics is the designing of FET parameters. An effective and precise approach is suggested in this study to determine the J112A n-Channel JFET's ideal parameters. PSO, FPA, and CSO, three distinct bio-inspired metaheuristic optimization techniques, were employed for the refinement, and the optimization process for the drain and transfer characteristics was the difference between the predicted and empirical quantity of the drain current. For the best model identification, three important criteria were chosen, namely RMSE of the training dataset, computational time, and convergence speed. In each case the calculated I-V characteristic is similar to the experimental I-V characteristic of n-channel JFET, which validates proposed methodology, but by means of RMSE, computational time and convergence speed favor FPA outcomes rather than PSO and CSO. Future parameter optimization methods for JFETs might be more effective and precise thanks to other cutting-edge and hybrid optimization techniques.

# References

1. Yan Y, Zhao Y, Liu Y. Recent progress in organic field-effect transistor-based integrated circuits. *Journal of Polymer Science* 2021.
2. Yuvaraja S, Nawaz A, Liu Q, Dubal D, Surya SG, Salama KN, Sonar P. Organic field-effect transistor-based flexible sensors. *Chemical Society Reviews* 2020;49: 3423–3460.
3. Nirantar S, Ahmed T, Ren G, Gutruf P, Xu C, Bhaskaran M, Walia S, Sriram S. Metal–air transistors: Semiconductor-free field-emission air-channel nanoelectronics. *Nano Venware* 2018;18: 7478–7484.
4. Cheng Z, Song X, Jiang L, Wang L, Sun J, Yang Z, Jian Y, Zhang S, Chen X, Zeng H. Mixed-dimensional WS2/GaSb heterojunction for high-performance pn diode and junction field-effect transistor. *Journal of Materials Chemistry C* 2022;10: 1511–1516.
5. Jazaeri F, Makris N, Saeidi A, Bucher M, Sallese J-M. Charge-based model for junction FETs. *IEEE Transactions on Electron Devices* 2018;65: 2694–2698.
6. Borghese A, Riccio M, Maresca L, et al. An experimentally verified 3.3 kV SiC MOSFET model suitable for high-current modules design. *Proceedings of the 2019 31st International Symposium on Power Semiconductor Devices and ICs (ISPSD)*. IEEE;2019, pp. 215–218.
7. Ezugwu AE, Shukla AK, Nath R, Akinyelu AA, Agushaka JO, Chiroma H Muhuri PK. Metaheuristics: A comprehensive overview and classification along with bibliometric analysis. *Artificial Intelligence Review* 2021;54: 1–80.
8. Fausto F, Reyna-Orta A, Cuevas E, AndradeÁG, Perez-Cisneros M. From ants to whales: Metaheuristics for all tastes. *Artificial Intelligence Review* 2020;53: 753–810.
9. Dutta P, Kumar A. Modelling of liquid flow control system using optimized genetic algorithm. *Statistics Optimization & Information Computing* 2020a;8: 565–582.
10. Dutta Pijush, Kumar A. Design an intelligent calibration technique using optimized GA-ANN for liquid flow control system. *J. Européen des Systèmes Automatisés* 2017a;50: 449–470. https://doi.org/10.3166/jesa.50.449-470
11. Dutta P, Majumder M, Kumar A. *Parametric optimization of liquid flow process by ANOVA optimized DE*. PSO & GA algorithms;2021b.
12. Dutta P, Agarwala R, Majumder M, Kumar A. Parameters extraction of a single diode solar cell model using bat algorithm, firefly algorithm & cuckoo search optimization. *Annals of Faculty Engineering Hunedoara* 2020;18: 147–156.
13. Dutta P, Majumder M, Kumar A. an improved grey wolf optimization algorithm for liquid flow control system. *International Journal of Engineering and Manufacturing* 2021c;4: 10–21.
14. Khan S, Bianchi T. Ant colony optimization (ACO) based data hiding in image complex region. *International Journal of Electrical and Computer Engineering* 2018;8: 379–389.
15. Kumar S, Solanki VK, Choudhary SK, Selamat A, González Crespo R. Comparative Study on Ant Colony Optimization (ACO) and K-Means Clustering Approaches for Jobs Scheduling and Energy Optimization Model in Internet of Things (IoT). *International Journal of Interactive Multimedia and Artificial Intelligence* 2020;6: 107–116.
16. Cai X, Wang H, Cui Z, Cai J, Xue Y, Wang L. Bat algorithm with triangle-flipping strategy for numerical optimization. *International Journal of Machine Learning and Cybernetics* 2018;9: 199–215.
17. Wang Y, Wang P, Zhang J, Cui Z, Cai X, Zhang W, Chen J. A novel bat algorithm with multiple strategies coupling for numerical optimization. *Mathematics* 2019;7: 135.
18. Dutta P, Biswas SK, Biswas S, Majumder M. Parametric optimization of solar parabolic collector using metaheuristic optimization. *The Computational Intelligence and Machine Learning* 2021a;2: 26–32.
19. Mandal S, Dutta P, Kumar A. Modeling of liquid flow control process using improved versions of elephant swarm water search algorithm. *SN Applied Sciences* 2019;1: 1–16.
20. Kumar V, Kumar D. A systematic review on firefly algorithm: Past, present, and future. *Archives of Computational Methods in Engineering* 2021;28: 3269–3291.

21. Wu J, Wang Y-G, Burrage K, Tian Y-C, Lawson B, Ding Z. An improved firefly algorithm for global continuous optimization problems. *Expert Systems with Applications* 2020;149: 113340.
22. Dokeroglu T, Sevinc E, Cosar A. Artificial bee colony optimization for the quadratic assignment problem. *Applied Soft Computing* 2019;76: 595–606.
23. Hancer E, Xue B, Zhang M, Karaboga D, Akay B. Pareto front feature selection based on artificial bee colony optimization. *Journal of Information Science* 2018;422: 462–479.
24. Dutta P, Kumar A. Modeling and optimization of a liquid flow process using an artificial neural network-based flower pollination algorithm. *Journal of Intelligent Systems* 2020b;29: 787–798.
25. Dutta P, Mandal S, Kumar A. Comparative study: FPA based response surface methodology & ANOVA for the parameter optimization in process control. *Advanced Modelling and Analysis C* 2018;73: 23–27. https://doi.org/10.18280/ama_c.730104
26. Mandal S, Dutta P, Kumar A. Application of FPA and ANOVA in the optimization of liquid flow control process. *Archives of Computational Methods in Engineering* 2018;5: 7–11. https://doi.org/10.18280/rces.050102
27. Wu G, ShenX, Li H, Chen H, Lin A, Suganthan PN. Ensemble of differential evolution variants. *Information Science* 2018;423: 172–186.
28. Alweshah M, Qadoura MA, Hammouri AI, Azmi MS, AlKhalaileh S. Flower pollination algorithm for solving classification problems. *International Journal of Advances in Soft Computing and its Applications* 2020;12: 15–34.
29. Abdel-Basset M, Shawky LA. Flower pollination algorithm: A comprehensive review. *Artificial Intelligence Review* 2019;52: 2533–2557.
30. Nguyen T-T, Pan J-S, Dao T-K. An improved flower pollination algorithm for optimizing layouts of nodes in wireless sensor network. *IEEE Access* 2019;7: 75985–75998.
31. Alyasseri ZAA, Khader AT, Al-Betar MA, Awadallah MA, Yang X-S. Variants of the flower pollination algorithm: A review. nat.-inspired algorithms. In *Nature-inspired algorithms and applied optimization*, Yang XS (Eds.). Springer;2018, Vol. 744, pp. 91–118.
32. Lei M, Zhou Y, Luo Q. Enhanced metaheuristic optimization: Wind-driven flower pollination algorithm. *IEEE Access* 2019;7: 111439–111465.
33. Dutta P, Kumar A. Application of an ANFIS model to optimize the liquid flow rate of a process control system. *Chemical Engineering Transactions* 2018a;71: 991–996.
34. Dutta P, Kumar A. Study of optimized NN model for liquid flow sensor based on different parameters. *International Conference on Materials, Applied Physics & Engineering (ICMAE 2018)*. 2018b.
35. Dutta P, Kumar A. Design an intelligent flow measurement technique by optimized fuzzy logic controller. *J. Européen des Systèmes Automatisés* 2018c;51: 89–107. https://doi.org/10.3166/jesa.51.89-107
36. Dutta P, Kumar A. Intelligent calibration technique using optimized fuzzy logic controller for ultrasonic flow sensor. *Mathematical Modelling of Engineering Problems* 2017b;4: 91–94. https://doi.org/10.18280/MMEP.040205

# 8

# AI-Based Model of Clinical and Epidemiological Factors for COVID-19

## 8.1 Introduction

The first case of pneumonia with unknown symptoms was found in December 2019 [1]. Since then, the virus epidemic has caused grave alarm, and on January 30, 2020, the World Health Organization (WHO) proclaimed the outbreak to be in full swing. Additionally, COVID-19, a novel coronavirus, was classified as a pandemic on March 11, 2020. Since there were 3,181,642 cases and 2,24,301 deaths impacting 215 nations and territories as of May 1, 2020, the novel coronavirus infection (COVID-19) caused shock and alarm throughout the world [2, 3]. At that time no effective vaccination or antiviral treatment was invented, so these numbers were anticipated to drastically increase in the future. The only effective method for halting the virus spread was preventative measures like lockdown. Accurate and timely information regarding the cause and epidemiological features of this worldwide pandemic may prove to be a viable means of containing the virus [4–6]. Finding the causes and spatial dissemination of these newly developing infectious diseases of zoonotic origin is extremely difficult due to their high transmissibility [8]. These complicated issues can be resolved with the effective integration of techniques like epidemiology, AI approaches, and bioinformatics. The Huanan Seafood Wholesale Market in Wuhan, where the first case of COVID-19 was discovered, was thought to be the primary site of the disease's propagation from animal to human. However, subsequent cases were not discovered to be related to the same exposure, and hence symptomatic human-to-human transmission was determined to be the primary factor in the spread of COVID-19. Presymptomatic and asymptomatic transmission for the disease's spread were both equally probable at the same time [7, 9]. Although a number of demographic parameters, such as gender, age, and blood type, have historically been categorized as infection risk factors, biostatistics studies have not yet developed sufficiently to explore complex correlations between numerous variables [10]. With the development of applications for neural network, fuzzy logic, and evolutionary algorithms, computers can now identify correlations between variables and determine which ones are most promising [11, 12].

In this research the authors mainly focus on the broad categorization of risk and preventive variables that contribute to the spread of COVID-19. To represent how the elements interact and are dependent on one another, different NN-model-based different function approaches are presented. Eight different training function based neural model are utilized for a given training dataset to verify which one of the training functions offered the regression value unity and the least mean square error (MSE) value is considered as a best fitted neural model for the present research. The manuscript is organized as follows: After the

DOI: 10.1201/9781003216001-10

introduction in Section 8.1, Section 8.2 covers the work related to the COVID-19 domain. Section 8.3.3 describes the modeling of neural network approach. Section 8.4 presents the result analysis and discussion of the proposed model. Section 8.5 presents the conclusion.

## 8.2 Related Work

The spread of the COVID-19 epidemic virus has captivated scientists and medical practitioners. To create effective preventive measures to stop the spread of COVID-19, it is necessary to comprehend the pattern of its unprecedented spread. Its limited span presents the biggest obstacle to fully comprehending the pattern of its proliferation. Therefore, much research is being done by scientists to understand its epidemic nature, which helps them predict the rise of infected persons with more accuracy. The recent studies pertaining to the forecast of its breakout are briefly covered in this section. Table 8.1 describes previous surveys conducted by several researchers.

**TABLE 8.1**

Survey of Previous Work

| Si No. | Reference | Survey |
|--------|-----------|--------|
| 1 | Li et al. [13] | Proposed a Gaussian distribution to explain the different stages of coronavirus transmission. Authors simulate the model for the same purpose while taking into account information on the Hubei epidemic situation. The authors assert that there is a relatively minor discrepancy between model predictions and actual values. The suggested approach can therefore serve as a foundation for epidemic prevention and control in the afflicted countries. |
| 2 | Okhuese [14] | The author observed the susceptible–exposed–infectious–removed (SEIR) model using migration data in this study until January 23, 2021. The author also used the most recent COVID-19 epidemiological data to comprehend the epidemic curve. Additionally, SEIR incorporates AI, and this improved SEIR model was trained using SARS data from 2003. In addition to this, the effectiveness of the model also verified for the forecasting of Ebola virus analysis in 2018. The author warrants that this dynamic SEIR model accurately forecasts the peaks and magnitude of the epidemic. |
| 3 | Al-Najjar and Al-Rousan [15] | Authors designed a neural network based classifier model to determine how a patient reacts to a therapy. The model was implemented by taking the dataset from February 20, 2020, to March 9, 2020, pertaining to recovered patients and patients who passed away. Seven different variables—country, area, infection reason, confirmation date, birth year, sex, and group—are used in the proposed model. The proposed classifier model looks for the characteristics that can most accurately predict death or recovery. |
| 4 | Cao et al. [16] | The authors applied knowledge based short-term preventative strategies that aid in the development of various control measures to stop the virus's further spread. For the same, they provide a time series model for COVID-19 short-term forecasting and a dynamic model for the epidemic. The forecasting was allegedly done accurately, according to the authors. |

*(Continued)*

**TABLE 8.1** (*Continued*)

| Si No. | Reference | Survey |
|--------|-----------|--------|
| 5 | Hellewell et al. [17] | Authors proposed a stochastic transmission mechanism to evaluate the efficacy of contact tracking and isolation, which led to this conclusion. This model calculates the number of primary infected individuals produced and the number of secondary infected individuals that contribute to calculating the reproduction number (R0). The infection rate increases exponentially as R0 rises over 1. The rate of infection could, however, exponentially decline if R0 stays below 1. |
| 6 | Sharma et al. [18] | Author employed data analytics to demonstrate how R0 can be managed by limiting interperson contact using a quarantine paradigm. This work also asserts that more stringent measures must be taken because the quarantine approach alone is insufficient to stop the spread of the disease. |
| 7 | Ranjan [19] | According to the author, the value R0 falls within India's predicted range of 1.4–3.9. Predictions are made using exponential and traditional susceptible–infected–recovered (SIR) models (both long-term and short-term). According to this model, equilibrium will be reached in India by May 2020. The authors also note that prediction is based on high degrees of social estrangement; it is invalid if communal transmission occurs in India. |
| 8 | Bannister-Tyrrell et al. [20] | In this article, the author took into account how the environment affected COVID-19's propagation and provided a fascinating statistic that mentioned the impact of temperature on its proliferation in Europe. According to the study, greater temperatures reduce the spread of this virus. |
| 9 | Lu et al. [21] | Author researched the connection between ventilation and epidemic. The scientists came to the conclusion that since droplets linger in the air longer in an air-conditioned environment, droplet transmission is improved. Additionally, the direction of air movement is a crucial consideration in this case. In addition to environmental influences, research tries to understand how physical health relates. |
| 10 | Rosita [22] | For the purpose of determining the prevalence of comorbidities in COVID-19 patients, the author presents an analysis based on a survey of several databases, including PubMed, EMBASE, and Web of Science. According to this study, severe patients are at an increased risk due to the current conditions compared to non-severe patients. |
| 11 | Dutta et al. [23] | Author designed a machine learning based model to predict whether a person is affected by COVID-19 on the basis of symptoms shown on his or her health status. Cold, fever, cough, body pain, and malaise were the most common potential symptoms for COVID. |
| 12 | Dutta et al. [24] | Author proposed four machine learning models to classify COVID-19 among other diseases like jaundice, malaria, covid, common cold, typhoid, dengue, and pneumonia based on feature selection and ranking methods. |

## 8.3 Artificial Neural Network Based Model

To classify the COVID-19 data using an ANN based model, a multilayer feed forward artificial neural n2etwork with one hidden layer which consist of 20 nodes has been used [25]. On the other hand, the input layer consists of 8 nodes which are different symptoms and features of COVID-19 patients. The output layer of the ANN model consists of only one node that indicates the predicted value (COVID 1 or 0), that is, the status of the patients. Initially, the training dataset is used to train the ANN model with the use of a back-propagation algorithm. The trained ANN model is then cross-validated using the training data itself, and the performance and accuracy was noted. Cross-validation is a resampling procedure used to evaluate machine learning models on a limited data sample [26]. Next, the trained model is tested against a new dataset that contained inputs for the period April 1 through April 7, 2020. The performance and accuracy are also noted for testing new cases. All the results are given in the results section.

### 8.3.1 Modeling of Artificial Neural Network

#### 8.3.1.1 Collection, Preprocessing, and Division of Data

For setting the artificial neural structure and implementation, we initially classify the test datasets from the clinical and epidemiological factors for the COVID-19 control framework [27, 28]. As the performance of neural network totally depends upon the full range of the input datasets, one must be take care while arranging the experimentation itself. A multilayer network is very useful for training data inside the range of inputs efficiently, but it doesn't have the caliber to train the data beyond this range. Input information of the neural system should be standardized before applying it. When the number of inputs is more than three, then the output of multilayer systems becomes substantially saturated as all the hidden layers used a tan-sigmoid activation function. In such cases, the gradient will be very small. System output consistently falls into a standardized range in the pre-preparing stage, and in the post-processing stage, system output follows the target output. To design a neural network model, input and target datasets are haphazardly divided into three different sets considered as a 90% training set and finally 10% information tests for testing set.

#### 8.3.1.2 Implementation of Neural Network

In the present research we used a dynamic and multilayer neural system for the model improvement of nonlinear clinical and epidemiological factors for COVID-19 strategy. In this proposal, we have utilized the feed-forward neural system design and derricks training algorithm mainly because of its simplicity of establishment. To implement this model, we utilize the neural system toolbox of MATLAB [29]. Initially to motivate the neural system model, the database is classified. A large number of datasets of the input parameters are used as the input row while risk value is the output row of a matrix. The output of the framework contains target datasets which have a straight relationship between risk value and input variables, namely virulence, immunity, temperature, populations, and ventilations.

In second steps, the NN model is found by considering the different transfer functions with the minimum mean square error (MSE) as shown in Table 8.2. Least MSE is better for the optimization. In MATLAB, the tool Tansig is available (second least MSE after Softmax), so in the present research we used transit as a neuron transfer function.

In feed-forward back-propagation, we used two hidden layers because from the tool analysis we get least MSE but maximum regression, which is shown in Table 8.3. To check the correlation between the output and target data in the NN model, we used regression analysis [29, 30]. Regression 1 indicates the close relationship between output and target data, and for 0 the relationship between output and target is random. Table 8.4 describes the details of the neural network model train, test, and validation datasets with the number of neurons and transfer function of each layer. Total details of input features and their ranges are shown in Table 8.5.

**TABLE 8.2**

List of Transfer Function with MSE

| SI. No | Transfer Function | MSE |
|--------|-------------------|-----|
| 1 | Linear Tanh | 2.21E-7 |
| 2 | Softmax | 0.99E-7 |
| 3 | Bias | 2.01E-7 |
| 4 | Linear | 3.66E-7 |
| 5 | Tanh | 2.62E-7 |
| 6 | Logsig | 1.23E-7 |
| 7 | Linear sigmoid | 2.54E-7 |
| 8 | Sigmoid | 2.47E-7 |
| 9 | Axon | 2.22E-7 |
| 10 | Tansig | 1.01E-7 |

**TABLE 8.3**

Feedforward Back-Propagation with MSE and R

| No. Hidden Layers | MSE | R |
|-------------------|-----|---|
| 1 | 9.6501 | 0.61672 |
| 2 | 1.63 | 0.99691 |
| 3 | 4.13 | 0.66979 |
| 4 | 6.838 | 0.98365 |
| 5 | 2.537 | 0.99142 |
| 6 | 4.936 | 0.98833 |

**TABLE 8.4**

Summary of the Optimized Neuron

| Database | Training Datasets | 144 |
|----------|-------------------|-----|
| | Test Datasets | 18 |
| No. of neurons in | 1st layer | 10 |
| | 2nd layer | 10 |
| Transfer function of | 1st layer | Tansig |
| | 2nd layer | Tansig |
| | Output layer | Linear |

**TABLE 8.5**

Input Range of ANN

| Input | Virulence | Immunity | Temperature | Population | Ventilation |
|-------|-----------|----------|-------------|-----------|-------------|
| Min | 4 | 5 | 15 | 300 | 18 |
| Max | 10 | 10 | 40 | 900 | 45 |

### 8.3.2  Performance of Training, Testing, and Validation of Network

In a neural system, the training process includes tuning the weights, biasing values and mean square error of the system with the assistance of successive epochs, and training function and input–output datasets. In this research we have utilized a back-propagation based training function so that neural systems could be accomplished over the whole length of the input space. In the training process, training datasets are applied for figuring out either the slope of the MSE or the Jacobian error, regarding the consecutive epochs, weights, and biases and improving this factor.

The validation datasets include MSE toward the finish of successive epochs. Like the training datasets MSE and gradient magnitude, the MSE and gradient magnitude of validation datasets were initially very small during the underlying stage [31–33]. However, when the training datasets are overfitted, training MSE may even now decrement despite the fact that validation MSE starts to prosper. If the validation MSE increases corresponding to the epochs, then the training process will be terminated and the model will obtain the weights and biases value during the least MSE of validation. In a specific epoch, if the slope size is not up to the limit set by the user, then the system updates the weights and biases to set least MSE toward the finish of the specific epoch. Aside from this, least training MSE, maximum training time, or extreme number of epochs are the other criteria to terminate the process [34]. In restricting criteria, training datasets also count test datasets' MSE in every epoch, but this is not used in a process terminating condition. If the test datasets accomplish a base value of MSE at an altogether unexpected epoch in comparison to the validation dataset's MSE, this shows poor segmentation of the datasets.

### 8.3.3  Performance Evaluation of Training Functions

In this segment, the presentation of various training functions utilized in feed-forward BP calculation, namely FG, BR, CGB, CGF, CGP, GD, GDM, GDA, GBX, LM, OSS, R, RP, and SCG, are assessed [35–38]. A sum of 162 exploratory examples are dealt with in the input–output target datasets for ANN-based clinical and epidemiological factors for COVID-19 model improvement. These ANN information tests have been randomly assigned such that the quantity of tests utilized for training and testing are 144, and 18, respectively.

Figure 8.1 shows the neural network model architecture applied in the present process, which consists of five inputs (virulence, immunity, temperature, populations, and ventilations) and one output (risk factor) parameter. In this architecture, the number of neurons in the output layer and hidden layer are 1 and 10, respectively. Figure 8.2 represents the regression plot for training, testing, and validation datasets of the NN model. For each of the cases the dataset regression (R) approaches to unity, which means the model is the best fit for predicting the datasets [39, 40]. Figure 8.3 shows the properties of the neural network based present model which indicates performance function, number of layer (hidden layer, output layer), their transfer function, the nature of the training function, and the adaptation learning function to predict the model accurately. Figure 8.2 represents the training information and training parameters like number of epochs, minimum gradation, maximum failure, and so on for the NN model.

Variation of the regression (R) and MSE with respect to number of nodes shown in Figures 8.3 and 8.4, respectively. From the graph it is seen that R approaches to unity and MSE is decreased by increasing the number of nodes of the present NN model.

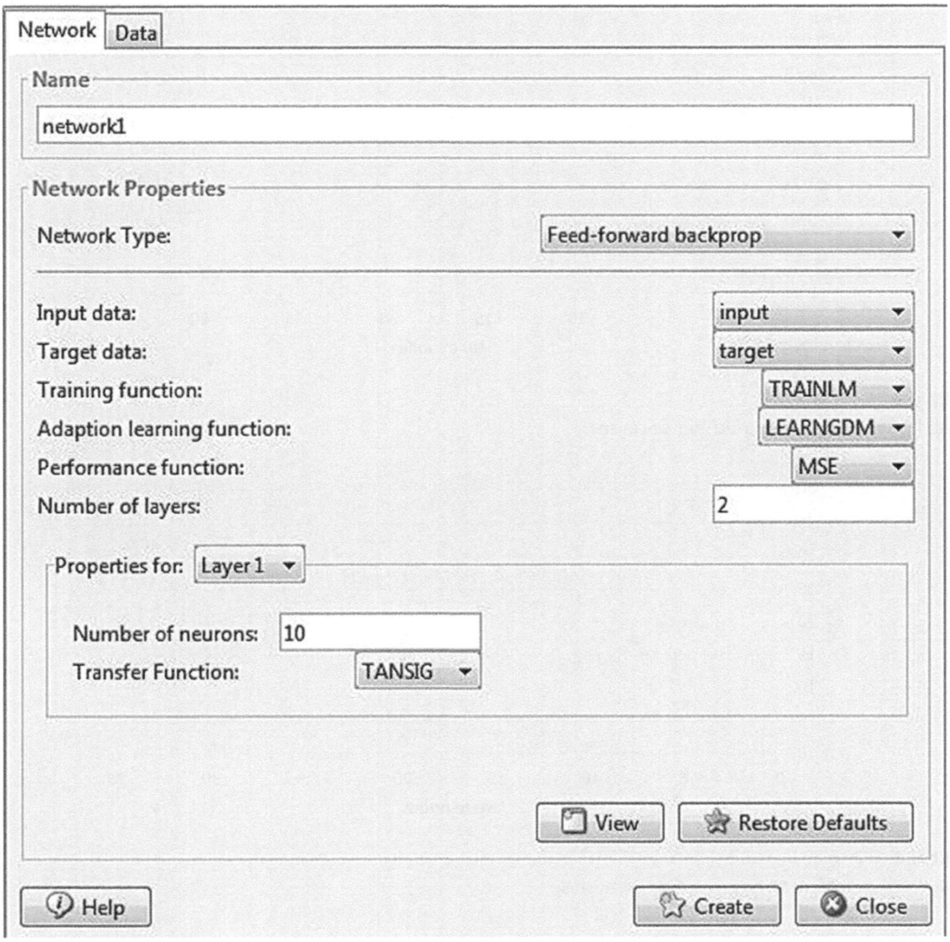

**FIGURE 8.1**
Properties of neural network diagram.

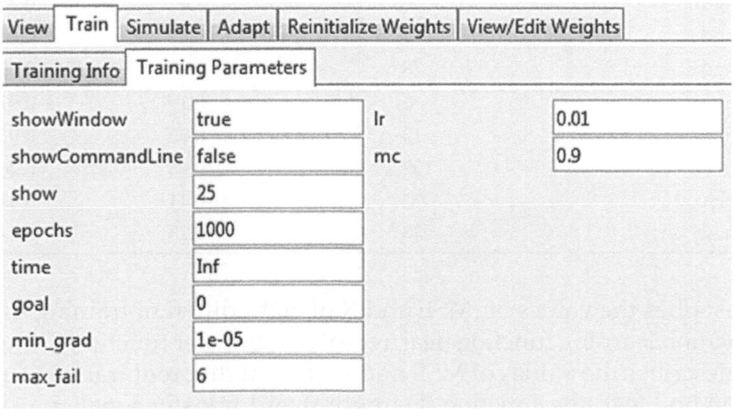

**FIGURE 8.2**
Training information and training parameters of neural network model.

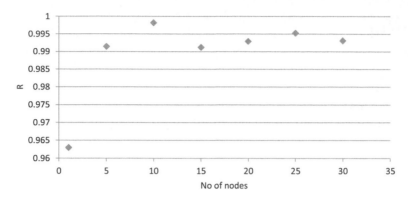

**FIGURE 8.3**
Plotting between R versus number of nodes.

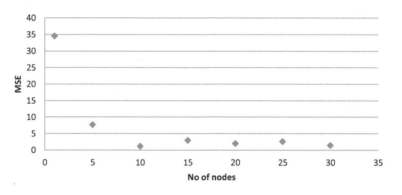

**FIGURE 8.4**
Plotting between MSE versus number of nodes.

**TABLE 8.6**

MSE and R for Purelin Transfer Function

| Training Function | No. of Iterations | MSE | R |
| --- | --- | --- | --- |
| Traingdm | 6 | 1.22 | 0.25134 |
| Traingda | 46 | 0.21 | 0.8434 |
| Traingdx | 28 | 0.41 | 0.9513 |
| Trainlm | 6 | 0.19 | 0.9734 |
| Trainoss | 13 | 0.35 | 0.9328 |
| Trainr | 157 | 0.95 | −0.2623 |
| Trainrp | 32 | 0.41 | 0.9314 |
| Trainscg | 28 | 0.28 | 0.9451 |

Table 8.6 describes the values of (MSE and R of eight different training functions under the same adaptation learning function (learngdm) and transfer function (Purelin). Similarly, Table 8.7 also describes the values of MSE and R of eight different training functions under the same adaptation learning function (learngdm) and transfer function (Tansig). In each of the cases it is seen that the Trainlm training function is best fitted for predicting the model.

**TABLE 8.7**

MSE and R for Tansig Transfer Function

| Training Function | No. of Iterations | MSE | R |
|---|---|---|---|
| Traingdm | 6 | 1.35 | 0.37648 |
| Traingda | 29 | 0.625 | 0.9804 |
| Traingdx | 28 | 2.37 | 0.06281 |
| Trainlm | 6 | 0.105 | 0.99818 |
| Trainoss | 13 | 5.57 | 0.5176 |
| Trainr | 157 | 11.95 | 0 |
| Trainrp | 23 | 0.903 | 0.9742 |
| Trainscg | 49 | 0.682 | 0.9915 |

## 8.4 Results and Discussion

In this section, the detailed results corresponding to two cases of AI models, namely, using ANN to detect if a person is affected by COVID or not, are shown and discussed. Figure 8.5 shows the neural network model (using MATLAB) to detect if any person is COVID positive or not based on some input symptoms and features. Table 8.8 and 8.9 present RMSE and accuracy of the proposed model for predicting the risk value with an accuracy of 95.77% and 98.02%, respectively.

After the training process is finished in the neural network model, the measurement framework is exposed to diverse input parameters. The output of the referenced process is portrayed Figures 8.5 and 8.6. Figure 8.5 presents the comparative study of actual risk

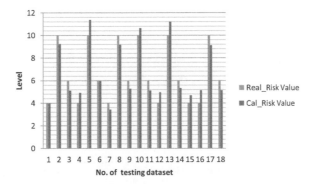

**FIGURE 8.5**

Plotting b real risk value versus calculated risk value.

**TABLE 8.8**

RMSE and Accuracy for Cross Validation

| Parameters | RMSE (%) | Accuracy (%) |
|---|---|---|
| Risk value | 4.23 | 95.77 |

**TABLE 8.9**

RMSE and Accuracy for Testing Dataset

| Parameters | RMSE (%) | Accuracy (%) |
|------------|----------|--------------|
| Risk value | 1.98     | 98.02        |

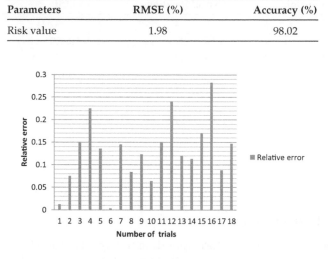

**FIGURE 8.6**
Relative error of risk value in test dataset versus number of trials.

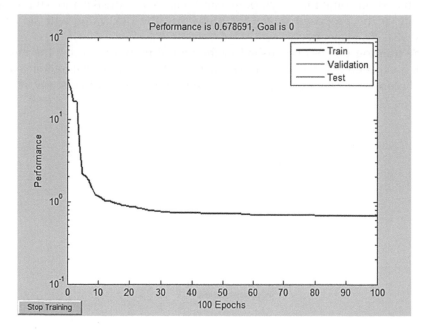

**FIGURE 8.7**
Plot for performance versus number of epochs.

value and calculated risk value for the testing dataset while Figure 8.6 presents relative error versus the number of testing trials dataset.

Figure 8.7 presents the performance versus of number of epochs for the present NN model. After proper tuning of the present model, the biasing weight of the NN layer is presented in Figure 8.8.

Select the weight or bias to view:   iw{1,1} – Weight to layer 1 from input 1   ▼

[0.55648 -15.5027 -22.4304 -21.6668 -11.3272;
 16.599 9.7002 -0.40516 7.7433 -2.8271;
 -27.828 10.3383 -9.5803 -13.2233 3.3604;
 -13.7664 8.325 -21.6583 -2.7315 -21.7751;
 -39.36 1.9726 0.90152 7.9454 -13.9186;
 25.7517 -26.1076 6.209 -28.9386 -5.7158;
 -25.9609 -21.0493 14.3667 -10.9316 22.3061;
 -9.121 10.5506 -12.0818 -29.2203 7.5286;
 17.993 10.2372 -2.7379 13.9955 -3.7221;
 0.51135 -0.14941 1.0152 -0.27535 1.4552]

Select the weight or bias to view:   lw{2,1} – Weight to layer   ▼

[0.20111 -0.41407 -0.24098 -0.57794 0.20093 -0.099183 0.18147 0.12844 0.6463 -1.1511]

Select the weight or bias to view:   b{1} – Bias to layer 1   ▼

[27.2706;
 -4.5154;
 22.5547;
 -24.6708;
 18.9076;
 -15.4033;
 -12.1092;
 -7.3792;
 -3.2601;
 1.3021]

Select the weight or bias to view:   b{2} – Bias to layer 2   ▼

[-0.2849]

**FIGURE 8.8**
Bias weighted value for the present NN model.

## 8.5 Conclusions

For this research, five different combinations of the algorithm along with 10 different transfer functions of the neurons are utilized. Among them, a feed-forward back-propagation NN model with Learngdm adaption learning function and Tansig transfer function is used due to the maximum value of R and minimum value of MSE, giving the maximum degree of accuracy. The model is designed with two hidden layers. The proposed optimization technique still fulfilled the objective for the cross validation and testing dataset with a higher degree of accuracy.

# References

1. Rahimi I, Chen F, Gandomi AH. A review on COVID-19 forecasting models. *Neural Computing and Applications* 2021: 1–11. https://doi.org/10.1007/s00521-020-05626-8

2. Sujath RAA, Chatterjee JM, Hassanien AE. A machine learning forecasting model for COVID-19 pandemic in India. *Stochastic Environmental Research and Risk Assessment* 2020;34(7): 959–972.

3. Jin X, Wang YX, Yan X. Inter-series attention model for covid-19 forecasting. *Proceedings of the 2021 SIAM International Conference on Data Mining (SDM)*. Society for Industrial and Applied Mathematics;2021, pp. 495–503.

4. Luo J. Forecasting COVID-19 pandemic: Unknown unknowns and predictive monitoring. *Technological forecasting and social change* 2021;166: 120602.

5. Huang CJ, Chen YH, Ma Y, Kuo PH. Multiple-input deep convolutional neural network model for covid-19 forecasting in china. *Med Rxiv* 2020.

6. Chimmula VKR, Zhang L. Time series forecasting of COVID-19 transmission in Canada using LSTM networks. *Chaos, Solitons & Fractals* 2020;135: 109864.

7. Petropoulos F, Makridakis S, Stylianou N. COVID-19: Forecasting confirmed cases and deaths with a simple time series model. *International Journal of Forecasting* 2020;38: 439–452.

8. Sarkar K, Khajanchi S, Nieto JJ. Modeling and forecasting the COVID-19 pandemic in India. *Chaos, Solitons & Fractals* 2020;139: 110049.

9. Vaishya R, Javaid M, Khan IH, Haleem A. Artificial intelligence (AI) applications for COVID-19 pandemic. *Diabetes & Metabolic Syndrome: Clinical Research & Reviews* 2020;14(4): 337–339.

10. Finucane MM, Samet JH, Horton NJ. Translational methods in biostatistics: Linear mixed effect regression models of alcohol consumption and HIV disease progression over time. *Epidemiologic Perspectives & Innovations* 2007;4(1): 1–14.

11. Tayfur G. *Soft computing in water resources engineering: Artificial neural networks, fuzzy logic and genetic algorithms*. WIT Press;2014.

12. Kim HJ, Shin KS. A hybrid approach based on neural networks and genetic algorithms for detecting temporal patterns in stock markets. *Applied Soft Computing* 2007;7(2): 569–576.

13. Li L, Yang Z, Dang Z, Meng C, Huang J, Meng H, … Shao Y. Propagation analysis and prediction of the COVID-19. *Infectious Disease Modelling* 2020;5: 282–292.

14. Okhuese AV. Estimation of the probability of reinfection with COVID-19 by the susceptible–exposed–infectious–removed–undetectable–susceptible model. *JMIR Public Health and Surveillance* 2020;6(2): e19097.

15. Al-Najjar H, Al-Rousan N. A classifier prediction model to predict the status of coronavirus COVID-19 patients in South Korea. *European Review for Medical and Pharmacological Sciences* 2020;24: 3400–3403.

16. Cao J, Jiang X, Zhao B. Mathematical modeling and epidemic prediction of COVID-19 and its significance to epidemic prevention and control measures. *Journal of Biomedical Research & Innovation* 2020;1(1): 1–19.

17. Hellewell J, Abbott S, Gimma A, Bosse NI, Jarvis CI, Russell TW, … Eggo RM (2020). Feasibility of controlling COVID-19 outbreaks by isolation of cases and contacts. *The Lancet Global Health* 8(4): e488–e496.

18. Sharma S, Volpert V, Banerjee M. Extended SEIQR type model for COVID-19 epidemic and data analysis. *MedRxiv* 2020.

19. Ranjan R. Predictions for COVID-19 outbreak in India using epidemiological models. *MedRxiv* 2020.

20. Bannister-Tyrrell M, Meyer A, Faverjon C, Cameron A. Preliminary evidence that higher temperatures are associated with lower incidence of COVID-19, for cases reported globally up to 29th February 2020. *MedRxiv* 2020.

21. Lu J, Gu J, Li K, Xu C, Su W, Lai Z, … Yang, Z. COVID-19 outbreak associated with air conditioning in restaurant, Guangzhou, China, 2020. *Emerging Infectious Diseases* 2020;26(7): 1628.

22. Rosita R. Pengaruh pandemi Covid-19 terhadap UMKM di Indonesia. *Jurnal Lentera Bisnis* 2020;9(2): 109–120.
23. Dutta P, Paul S, Kumar A. Comparative analysis of various supervised machine learning techniques for diagnosis of COVID-19. In *Electronic devices, circuits, and systems for biomedical applications*, Tripathi S, Prakash K, Balas V, Mohapatra S, Nayak J (Eds.). Academic Press;2021, pp. 521–540.
24. Dutta P, Paul S, Obaid AJ, Pal S, Mukhopadhyay K. Feature selection based artificial intelligence techniques for the prediction of COVID like diseases. In *Journal of Physics: Conference Series*. IOP Publishing;2021, July, Vol. 1963, No. 1, p. 012167.
25. Abiodun OI, Jantan A, Omolara AE, Dada KV, Mohamed NA, Arshad H. State-of-the-art in artificial neural network applications: A survey. *Heliyon* 2018;4(11): e00938.
26. Wang SC. Artificial neural network. In *Interdisciplinary computing in java programming* Springer;2003, pp. 81–100.
27. Vohradsky J. Neural network model of gene expression. *The FASEB Journal* 2001;15(3): 846–854.
28. Eberhart RC (Ed.). *Neural network PC tools: A practical guide*. Academic Press;2014.
29. Dayhoff JE. *Neural network architectures: An introduction*. Van Nostrand Reinhold Co.;1990.
30. Hecht-Nielsen R. Theory of the backpropagation neural network. In *Neural Networks for Perception*. Academic Press;1992, pp. 65–93.
31. Gardner E. The space of interactions in neural network models. *Journal of Physics A: Mathematical and General* 1988;21(1): 257.
32. Gallant SI, Gallant SI. *Neural network learning and expert systems*. MIT press;1993.
33. Beale MH, HaganMT, Demuth HB. Neural network toolbox. *User's Guide, MathWorks* 2010;2: 77–81.
34. Anthony M, Bartlett PL, Bartlett PL. *Neural network learning: Theoretical foundations*. Cambridge University Press;1999, Vol. 9.
35. Du KL. Clustering: A neural network approach. *Neural Networks* 2010;23(1): 89–107.
36. Psaltis D, Sideris A, Yamamura AA. A multilayered neural network controller. *IEEE Control Systems Magazine* 1988;8(2): 17–21.
37. Ding S, Su C, Yu J. An optimizing BP neural network algorithm based on genetic algorithm. *Artificial Intelligence Review* 2011;36(2): 153–162.
38. Lin CT, Lee CSG. Neural-network-based fuzzy logic control and decision system. *IEEE Transactions on Computers* 1991;40(12): 1320–1336.
39. Abiodun OI, Jantan A, Omolara AE, Dada KV, Umar AM, Linus OU, … Kiru MU. Comprehensive review of artificial neural network applications to pattern recognition. *IEEE Access* 2019;7: 158820–158846.
40. Elsheikh AH, Sharshir SW, Abd Elaziz M, Kabeel AE, Guilan W, Haiou Z. Modeling of solar energy systems using artificial neural network: A comprehensive review. *Solar Energy* 2019;180: 622–639.

# 9

## Fuzzy Logic Based Parametric Optimization Technique of Electro Chemical Discharge Micro-Machining (µ-CDM) Process during Micro-Channel Cutting on Silica Glass

### 9.1 Introduction

The necessity for complex micro-miniature materials with better technical precision has grown over the years as a result of technological advancements in both conductive and nonconductive materials. On the other hand, nonconductive materials, in particular silica, are widely employed for miniature process. Engineers, however, face a difficult problem in machining these materials with high dimensional accuracy. Due to their intrinsic hardness and brittleness restrictions, traditional micromachining techniques like micro-drilling, micro-milling, and micro-grinding cannot fabricate complex micro features within the tolerance level.

Multiple nontraditional machining approaches have been used to solve problems with conventional machining and have successfully machined regardless of their mechanical and chemical properties, molding the materials into the necessary forms with surface quality adherence. The most widely used and marketed procedures for advanced micro-manufacturing are micro-electro chemical machining (µ-ECM) and micro-electro discharge machining (µ-EDM). However, there are several issues with µ-ECDM that are inherent, such as poor surface quality, a decreased material removal rate (MRR), a high rate of tool wear, thermally driven cracks and the production of recast layers.

There are several research work was done for the fabrication of miniature parts using electrochemical discharge micro-machining process (µ-ECDM). Singh and Singh [1] illustrated the influence of tool feed rate, tool rotation, and duty cycle on material removal rate and overcut of a machined hole and optimized those machining criteria using a novel combined entropy-VIKOR method. A Taguchi's methodology used electrolyte concentration, applied voltage, and inter-electrode gap as process parameters on the output characteristics of material removal rate (MRR) and overcut rate. A multiobjective process optimization grey relational analysis (GRA) method is used for achieving the optimum response variables [2]. Mallick et al. [3] explore the effects of various process parameters of µ-ECDM process based on the relationship between the machining parameters such as electrolyte concentration (wt%), applied voltage ($V$), width of cut (WOC), surface roughness (Ra), tool shapes on heat affected zone (HAZ) and material removal rate (MRR). A comparative study based on RSM-based GA and RM-based PSO is performed for

DOI: 10.1201/9781003216001-11

obtained optimum process parameters in an ECDM process [4]. A multi-response process optimization TOPSIS is used for the analysis of response variables of overcut and material removal rate (MRR) corresponding to input process parameters of electrolyte concentration gap, applied voltage, and inter-electrode gap [1]. A single as well as multiobjective optimization is applied using a genetic algorithm (GA) for obtaining the suitable optimal parameters in an ECDM process [5]. It was observed that moderate overcut and highest MRR are obtained for suspended electrolyte with stirrer effect [6]. In Chen et al. [7] a proper ultrasonic amplitude is added, and it is found that machining long-term performance is greatly enhanced. In order to control the operative gap and stable gas films directly below the micro-tool, Singh and Dvivedi [8] employed a pressurized feeding system and abrasive-coated tooling. According to Wang et al. [9], as the DC voltage increased, the MRR and surface resilience first increased and declined slightly. The bending force was delivered to a micro-tool in the perspective of a magnetism in order to show [10] that it decreased as the electrolytic concentrations and applied voltage raised. Han et al. [11] enhanced step milling depth and produced micro-grooves with a high aspect ratio using the ECDM technique. We investigated the effects of control parameters on silicon carbide during micro-drilling [12]. By combining sodium hydroxide (NaOH) and potassium hydroxide (KOH), Sabahi and Razfar [13] increased the accuracy of performances of electrochemical discharge machining (ECDM). Yadav [14] examined and recorded the prospects for electrochemical spark machining (ECSM) research in the current state of the field and created intricate profiles with improved surface quality. Tang et al. [15] examined the effects of current pulses on the gas film and found that, lead to a full gas film developing, a sizable bubble was created around the electrode as a result of gas generation and bubble flocculation. The gas film was migrating upward at a mean speed of 1.03 m/s, according to the study. Oza et al. [16] employed Taguchi resilient design and L9 orthogonal array to identify the most appropriate parameterization conditions for surface texture and kerf width qualities. The tool for the travelling wire electrochemical discharge machining was wrapped wire with a 0.15 mm diameter, according to the experts. Bindu Madhavi and Hiremath [17] used a 370 m diameter stainless steel (SS) tool to construct a micro-channel on 4 mm thick quartz glass utilising ECDM while taking electrolyte concentration (wt% C), varying voltage (V), and duty factor levels into consideration (percent DF). Bellubbi and Mallick [18] provided an illustration showing how surface imperfections increase as stand-off distance rises. In present research fuzzy logic based AI techniques are used for identification of response variables corresponding to input variable as well as predict the optimum response variable for a new set of testing data. There have been several researches performed on fuzzy logic such as the measured intelligent flow measurement technique [19, 20], turbidity measurement [21], performance analysis of flow sensor [22]. An improved version of elephant swarm optimization (ESWSA) algorithm is applied to get the optimum input parameters on an Aluminum 6061T6 plate [23].

As a result, it is evident from the literature review that various researchers and scientists' efforts have been made to accomplish their objectives, however a key focus might be placed on the area of micro-profile or micro-channel cutting on electrically insulating materials, such glass, employing the ECDM technique to increase machining depth by merging mechanized spring feed operation and Z axis motor. In this research article paper organized in the following steps, in the first stage a micro ECDM set up is analyzed in Section 9.2, methodology and experimental results are explained Section 9.3, and fuzzy model implementation followed by result analysis and conclusion is described in Section 9.4.

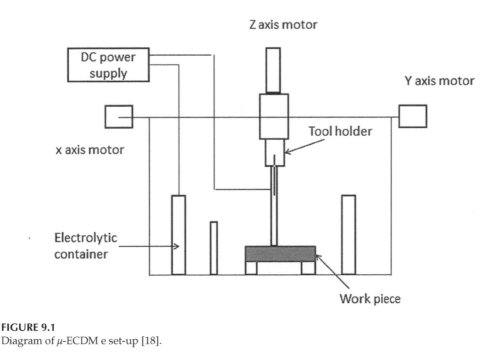

**FIGURE 9.1**
Diagram of $\mu$-ECDM e set-up [18].

## 9.2 Development of the Set Up

To begin pursuing the goals of this study project and manage process variables like pulse on time (PT), machining voltage ($v$), and electrolyte concentration (EC), a homegrown design and development was made for an experimental ECDM setup. The experimental set-up for ECDM is depicted in Figure 9.1 and includes an automated spring feed system for tool movement and fixation along the X-Y-Z axis.

## 9.3 Experimental Methodology and Result Analysis

The data input parameters are electrolyte concentration (EC) (10,17.5, 25 wt%), pulse on time (PT) (40,50,60 µs), applied voltage ($v$) (45,50,55 V), stand-off distance (SD)(0.5,1,1.5 mm) and the average values of the experimental findings so achieved, each experiment being carried out three times under the identical machining circumstances with the defined specifications of Tungsten tool tip diameter 200 µm and inter-electrode gap (IEG) 30 mm, pulse frequency 50 Hz. Slip gauges were inserted between the work and a tungsten tool tip to measure the stand-off distance. Table 9.1 shows the empirical design and test outcomes. To perform this study, out of 27 datasets, 24 datasets were used to construct the fuzzy logic model implemented by four input variables and three output variables. Three testing data-sets were used for cross validation purposes.

**TABLE 9.1**

Training and Testing Experimental Dataset [18]

| Run Order | Voltage (V) (x1) | Pulse on Time (µs) (x2) | Stand-off Distance (mm) (x3) | Electrolyte Concentration (wt%) (x4) | MRR (mg/hr) | OC (µm) | MD (µm) |
|---|---|---|---|---|---|---|---|
| 1 | 45 | 40 | 1 | 17.5 | 72.93 | 172.328 | 291.722 |
| 2 | 55 | 40 | 1 | 17.5 | 88.65 | 263.234 | 354.65 |
| 3 | 45 | 60 | 1 | 17.5 | 86.413 | 186.43 | 345.642 |
| 4 | 55 | 60 | 1 | 17.5 | 102.42 | 301.1 | 409.678 |
| 5 | 50 | 50 | 0.5 | 10 | 77.56 | 210.21 | 310.25 |
| 6 | 50 | 50 | 1.5 | 10 | 56.87 | 246.797 | 220.48 |
| 7 | 50 | 50 | 0.5 | 25 | 83.96 | 155.24 | 335.86 |
| 8 | 50 | 50 | 1.5 | 25 | 82.24 | 226.447 | 328.957 |
| 9 | 45 | 50 | 1 | 10 | 45.55 | 118.365 | 182.2 |
| 10 | 55 | 50 | 1 | 10 | 90.445 | 225.1 | 361.82 |
| 11 | 45 | 50 | 1 | 25 | 88.45 | 106.356 | 353.83 |
| 12 | 55 | 50 | 1 | 25 | 76.56 | 310.12 | 306.24 |
| 13 | 50 | 40 | 0.5 | 17.5 | 84.34 | 154.856 | 337.37 |
| 14 | 50 | 60 | 0.5 | 17.5 | 98.22 | 227.494 | 392.88 |
| 15 | 50 | 40 | 1.5 | 17.5 | 80.38 | 225.23 | 321.52 |
| 16 | 50 | 60 | 1.5 | 17.5 | 91.15 | 319.584 | 364.54 |
| 17 | 45 | 50 | 0.5 | 17.5 | 77.47 | 102.299 | 309.89 |
| 18 | 55 | 50 | 0.5 | 17.5 | 90.844 | 292.867 | 363.376 |
| 19 | 45 | 50 | 1.5 | 17.5 | 68.42 | 213.223 | 273.78 |
| 20 | 55 | 50 | 1.5 | 17.5 | 84.36 | 294.494 | 337.33 |
| 21 | 50 | 40 | 1 | 10 | 56.44 | 170.265 | 225.76 |
| 22 | 50 | 60 | 1 | 10 | 82.98 | 208 | 331.92 |
| 23 | 50 | 40 | 1 | 25 | 87.32 | 248.376 | 349.23 |
| 24 | 50 | 60 | 1 | 25 | 85.24 | 269.243 | 340.97 |
| 25 | 50 | 50 | 1 | 17.5 | 67.18 | 244.212 | 268.75 |
| 26 | 50 | 50 | 1 | 17.5 | 66.12 | 247.278 | 264.48 |
| 27 | 50 | 50 | 1 | 17.5 | 65.82 | 242.278 | 263.39 |

### 9.3.1 Effects of Process Parameters on MRR, OC, and MD

Figure 9.2 represents the fuzzy inference system for the present model where four input effects were pulse-on time (PT), electrolyte concentration (EC), applied voltage (v), and stand-off distance (SD), and the effects on the response variables were OC, MRR, and MD.

Figures 9.3–9.6 present the triangular membership function of input variables pulse-on time (PT), applied voltage (v), electrolyte concentration (EC), and stand-off distance (SD), where each of the input variables has three membership functions possible, creating the combination of 81. Figures 9.7–9.9 present the membership function of the output variables OC, MRR, and MD. Figure 9.10 indicates the possible logic statements between four input

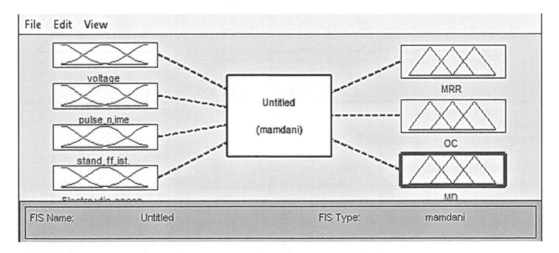

**FIGURE 9.2**
FIS model for input and output process variables.

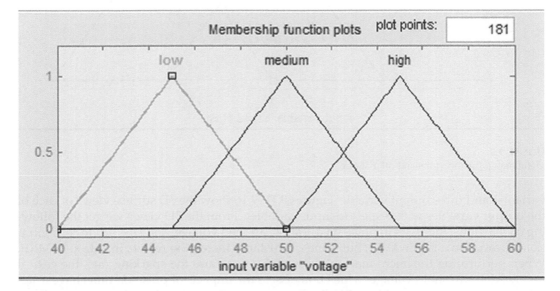

**FIGURE 9.3**
Membership function for voltage.

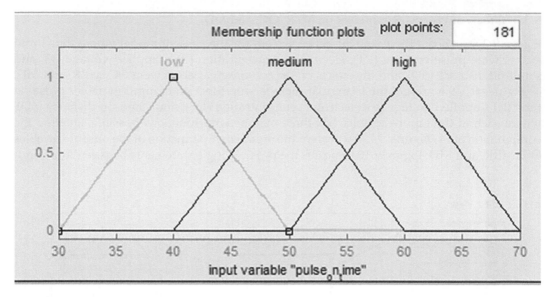

**FIGURE 9.4**
Membership function for pulse on time.

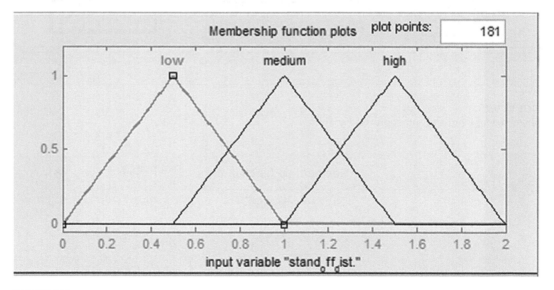

**FIGURE 9.5**
Membership function for stand-off distance.

variables and three output variable. Figures 9.11–9.16 show the 3D surface views of each of the output variables with respect to input variables. From the 3D views we get the following conclusions. Increases in pulse-on duration, applied voltage, and electrolyte concentration all result in a rise in MRR, but stand-off distance increasing results in a drop in MRR. When performing the micro-machining procedures to raise the sparking rate, the critical voltage and threshold voltage, which both assess the impacts of pulse-on time, have a significant impact on the machinability. Two parameters are modified for the purpose of analyzing the parametric effects on the machining criteria, while the remaining parameters are

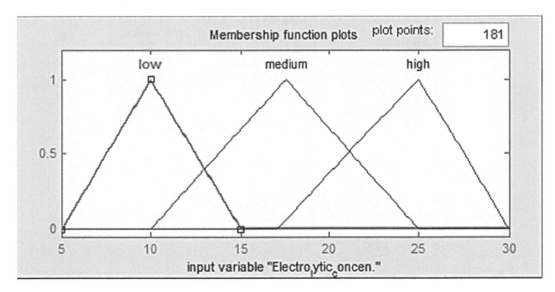

**FIGURE 9.6**
Membership function for electrolytic concentration.

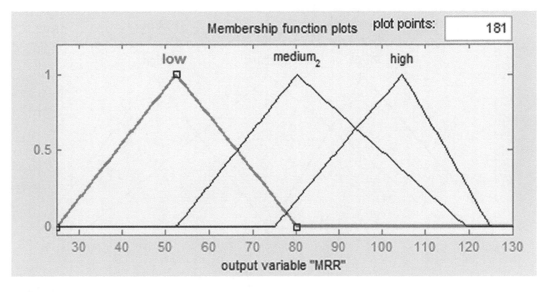

**FIGURE 9.7**
Membership function for MRR.

held at their accurate estimation and all fixed parameters are maintained constant. Increases in pulse-on duration, applied voltage, and electrolyte concentration all result in an increase in operating current (OC), whereas drops in stand-off range result from slowing spark rates. Side sparking and OC are both enhanced when the stand-off distance is enhanced. It is also evident that when sparking rate is raised, applied voltage and pulse-on time rise. At 50 V, consistent sparking is produced, which improves surface properties. Electrolyte density of 17.5 wt% also accompanies and lowers if stand-off distance (SD) is enhanced because the sparking rate is lessened.

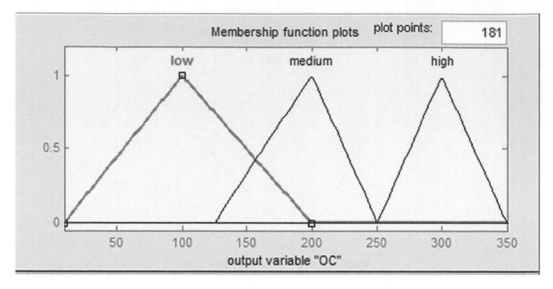

**FIGURE 9.8**
Membership function for OC.

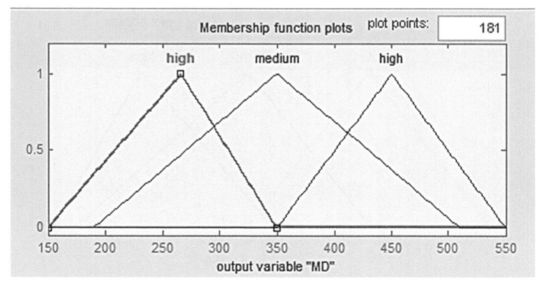

**FIGURE 9.9**
Membership function for MD.

### 9.3.2 Determination of Optimized Condition

Figures 9.11–9.16 show the parametric combination for minimum OC, maximum MRR, and maximum MD. From Figure 9.11 and 9.14 it is found that maximum MRR is tracked at the maximum level of applied voltage, pulse-on time, medium electrolytic concentration, stand-off distance. From Figure 9.12 and 9.15 it is added that for minimum supply voltage, maximum electrolytic concentration, and medium stand-off distance and pulse-on time, OC is minimum. From Figure 9.13 and 9.16 it is found that for maximum pulse-on time, and medium voltage, electrolytic concentration, and stand-off distance, MD is maximum.

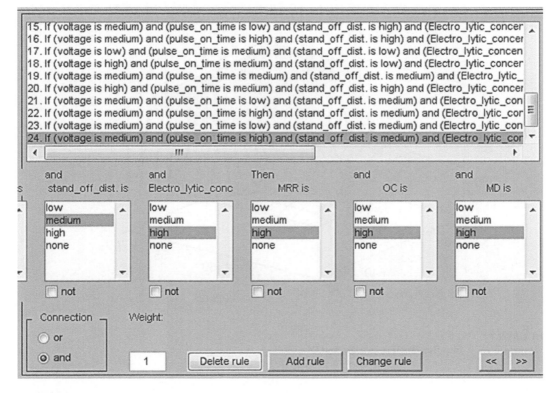

**FIGURE 9.10**
Fuzzy inference rule between process input and output variables.

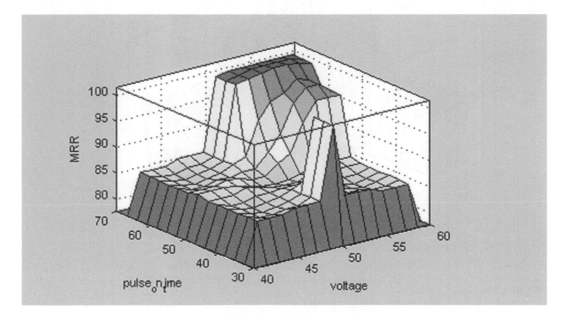

**FIGURE 9.11**
3D surface view for voltage, pulse-on time, and MRR.

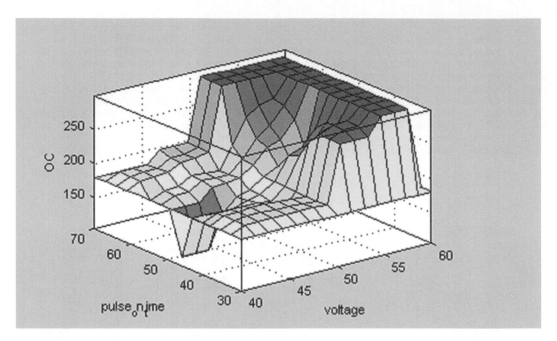

**FIGURE 9.12**
3D surface view for pulse-on time, voltage, and OC.

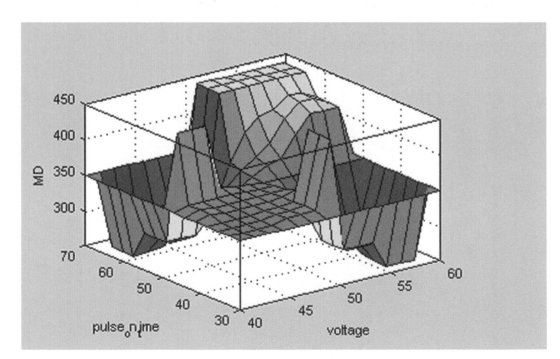

**FIGURE 9.13**
3D surface view for pulse-on time, voltage, and MD.

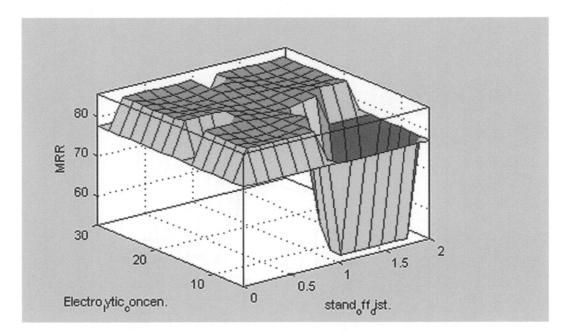

**FIGURE 9.14**
3D surface view for stand-off distance, electrolytic concentration, and MRR.

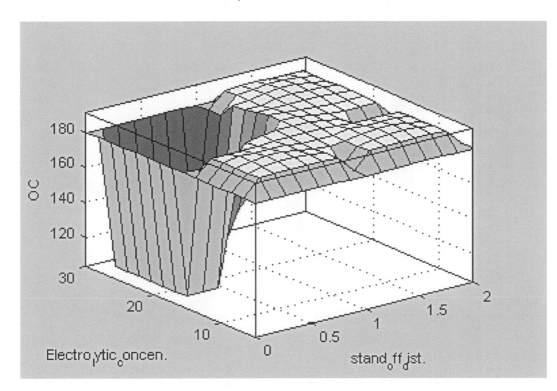

**FIGURE 9.15**
3D surface view of electrolytic concentration, stand-off distance, and OC.

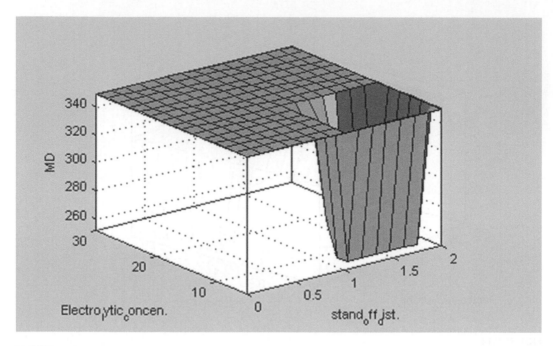

**FIGURE 9.16**
3D surface view of electrolytic concentration, stand-off distance, and MD.

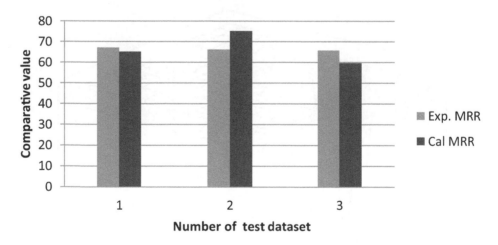

**FIGURE 9.17**
Experimental and calculated value of test dataset for MRR.

Figures 9.17–9.19 present the comparative study of response variables MRR, OC, and MD for the test dataset. For each of the characteristics graphs it has been seen that both the calculated results follow the experimental results. From Table 9.2 it has been concluded that the RMSE error of the response variable is 3.68%, 5.64%, and 6.45% for MRR, OC, and MD, respectively.

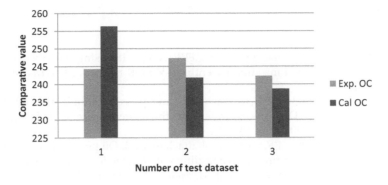

**FIGURE 9.18**
Experimental and calculated value of test dataset for OC.

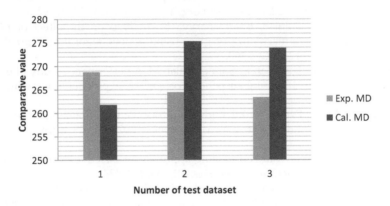

**FIGURE 9.19**
Experimental and calculated value of test dataset for MD.

**TABLE 9.2**

RMSE for the Test Dataset

| Name of Response Variable | MRR (%) | OC (%) | MD (%) |
| --- | --- | --- | --- |
| RMSE | 3.68 | 5.64 | 6.45 |

## 9.4 Conclusions

The accompanying information was gathered from the parametric study of the existing study's capabilities of the channel cutting on silica glass while using an ECDM setup. This analysis was based on extensive experimental observations and desire function analysis. The overall research segments into two parts; in the first part optimal process parameters are achieved corresponding to independent input variables already discussed in the previous section while in the second section prediction of the testing data set has been performed. Regarding these two research outputs, the following conclusion has been made. By raising voltage, electrolyte concentration, and pulse-on time, the machining depth, MRR, and OC rise; however, when stand-off distance is raised, these parameters drop.

## References

1. Singh M, Singh S. Micro-machining of CFRP composite using electrochemical discharge machining and process optimization by entropy-VIKOR method. *Materials Today: Proceedings* 2021;44: 260–265.

2. Garg MP, Singh M, Singh S. Micro-machining and process optimization of electrochemical discharge machining (ECDM) process by GRA method. *International scientific-technical conference manufacturing*. Springer;2019, pp. 384–392.

3. Mallick B, Biswas S, Sarkar BR, Doloi B, Bhattacharyya B. On performance of electrochemical discharge micro-machining process using different electrolytes and tool shapes. *International Journal of Manufacturing, Materials, and Mechanical Engineering* 2020;10: 49–63. https://doi.org/10.4018/IJMMME.2020040103

4. Naik R, Sathisha N. Desirability function and GA-PSO based optimization of electrochemical discharge micro-machining performances during micro-channeling on silicon-wafer using mixed electrolyte. *Silicon* 2022. https://doi.org/10.1007/s12633-022-01697-5

5. Mallick B, Sarkar BR, Doloi B, Bhattacharyya B. Analysis on the effect of ECDM process parameters during micro-machining of glass using genetic algorithm. *Journal of Mechanical Engineering Science* 2018;12: 3942–3960.

6. Vinod Kumaar JR, Thanigaivelan R, Soundarrajan M. A performance study of electrochemical micro-machining on SS 316L using suspended copper metal powder along with stirring effect. *Materials and Manufacturing Processes* 2022;37: 1–14.

7. Chen Y, Feng X, Xin G. Experimental study on ultrasonic vibration-assisted WECDM of glass microstructures with a high aspect ratio. *Micromachines* 2021;12: 125. https://doi.org/10.3390/mi12020125

8. Singh T, Dvivedi A. On pressurized feeding approach for effective control on working gap in ECDM. *Materials and Manufacturing Processes* 2018;33: 462–473.

9. Wang J, Sun L, Jia Z. Research on electrochemical discharge-assisted diamond wire cutting of insulating ceramics. *International Journal of Advanced Manufacturing Technology* 2017;93: 3043–3051.

10. Hajian M, Razfar MR, Etefagh AH. Experimental study of tool bending force and feed rate in ECDM milling. *International Journal of Advanced Manufacturing Technology* 2017;91: 1677–1687.

11. Han M-S, Chae KW, Min B-K. Fabrication of high-aspect-ratio microgrooves using an electrochemical discharge micromilling process. *Journal of Micromechanics and Microengineering* 2017;27: 055004.

12. Sarkar BR, Doloi B, Bhattacharyya B. Experimental investigation into electrochemical discharge microdrilling on advanced ceramics. *International Journal of Manufacturing Technology and Management* 2008;13: 214–225.

13. Sabahi N, Razfar MR. Investigating the effect of mixed alkaline electrolyte (NaOH+ KOH) on the improvement of machining efficiency in 2D electrochemical discharge machining (ECDM). *International Journal of Advanced Manufacturing Technology* 2018;95: 643–657.

14. Yadav RN. Electro-chemical spark machining–based hybrid machining processes: Research trends and opportunities. *Proceedings of the Institution of Mechanical Engineers - Part B: Journal of Engineering Manufacture* 2019;233: 1037–1061.

15. Tang W, Kang X, Zhao W. Experimental investigation of gas evolution in electrochemical discharge machining process. *International Journal of Electrochemical Science* 2019;14: 970–984.

16. Oza AD, Kumar A, Badheka V, Arora A. Traveling wire electrochemical discharge machining (TW-ECDM) of quartz using zinc coated brass wire: Investigations on material removal rate and kerf width characteristics. *Silicon* 2019;11: 2873–2884.

17. Bindu Madhavi J, Hiremath SS. Machining and characterization of channels and textures on quartz glass using µ-ECDM process. *Silicon* 2019;11: 2919–2931.

18. Bellubbi S, Mallick B. Multi response optimization of ECDM process parameters for machining of microchannel in silica glass using Taguchi–GRA technique. *Silicon* 2022;14: 4249–4263.

19. Dutta P, Kumar A. Intelligent calibration technique using optimized fuzzy logic controller for ultrasonic flow sensor. *Mathematical Modelling of Engineering Problems* 2017a;4: 91–94. https://doi.org/10.18280/mmep.040205
20. Dutta P, Kumar A. Design an intelligent flow measurement technique by optimized fuzzy logic controller. *Journal Europeen des Systemes Automatises* 2018;51: 89–107.
21. Dutta P, Kumar A. Fuzzy model for tubidity measurement. *International Journal of Advanced Computer Technology* 2015;4: 41–45.
22. Dutta P, Kumar A. Effect of different defuzzification methods in a fuzzy based liquid flow control in semiconductor based anemometer. *International Journal of Information Technology, Control and Automation* 2017b;7. https://doi.org/10.5121/ijitca.2017.7101
23. Dutta P, Majumder M. Parametric optimization of drilling parameters in aluminum 6061t6 plate to minimize the burr. *International Journal of Engineering and Manufacturing* 2021;11: 36–47. https://doi.org/10.5815/ijem.2021.06.04

# 10

## Study of ANFIS Model to Forecast the Average Localization Error (ALE) with Applications to Wireless Sensor Networks (WSN)

### 10.1 Introduction

A wireless sensor network (WSN) is a self-organizing network made up of numerous small, inexpensive sensor nodes that can track changes in the physical or environmental characteristics [1–3]. Its practical applications include energy harvesting [4], health monitoring [5], precision farming [6], target tracking [7], transportation management [8], global-scale wildlife monitoring [9], environmental monitoring [10], and business and home automation [9, 11]. The majority of applications call for these sensors to estimate their coordinates precisely while using fewer resources. These sensors include an inbuilt Global Positioning System (GPS) device that allows them to find their coordinates quickly [12]. Due to its size and expense, GPS cannot be feasibly incorporated into all sensors. A number of algorithms have been applied for the localization of unknown nodes with the help of anchor nodes [13, 14].

To address various localization issues, numerous localization methods have been developed [15]. These algorithms must be adaptable in order to function successfully in a wide range of indoor and outdoor settings and topologies. The four types of localization protocol used in WSN are range-free algorithms, range-based algorithms, centralized algorithms, and decentralized algorithms. In the range-based approach, nodes choose their places by calculating the angular distance from anchor nodes. There are several parametric techniques like time of arrival (ToA), angle of arrival (AoA), time difference of arrival (TDoA), and received signal strength indicator (RSSI) used to acquire these estimations [16–18].

Due the limitations of sensor-based equipment, range-free localization techniques are a financially distinct choice. Range-free localization adopts two different protocols, namely (1) centroid methods, where a high density of sensor node sites get historic locations in the forms of range-free localization protocols [19, 20] and (2) hop-based techniques, which rely on saturating the network with connectivity data like hop count [21, 22].

The data transmission to a central node is necessary for the centralized approach in order to compute the location of the mobile node. Due to the power limitations on each sensor as well as the lengthy multihop information transmission, this technique is fairly expensive. As a result, any connection to a facility for centralized computing is expensive because each sensor node has limited power accessibility [23, 24]. Additionally, delivering time arrangement information over a network introduces latency and also uses more energy and network bandwidth. Decentralized localization techniques, on the other hand,

DOI: 10.1201/9781003216001-12

necessitate fewer connections between sensor nodes, which reduce the WSN's power usage. Decentralized localization frameworks need hardware based components to make a connection with each portable target in order to collect localization data from reference nodes [25, 26].

In recent years, AI techniques have become more and more important in engineering optimization [27–35]. This can be explained by the fact that they perform better than traditional mathematical optimization methods while using significantly less computing power and memory. Modeling WSN node localization typically involves solving a multidimensional optimization issue. Various AI techniques have been presented for range-based approaches in an effort to design a less complex algorithm [36]. A two-step AI model is used for perfect localization in a WSN where a range-free localization strategy based on received signal intensity has been applied. In the first approach, nodes are localized using a fuzzy logic system (FLS) by adding the edge weights of each anchor node. Then, a genetic algorithm is used to calculate the ideal edge weight (GA). The second approach really employs a neural network (NN) technique, where the output is the approximate position of wireless nodes and the input is the received signal intensity. Utilizing FLS and GA, simulation results using NN are in contrast [34]. In [37, 38], the authors suggested a weighted centroid localization technique-based "range-free" localization strategy for WSNs. During the construction of membership function for RSSI and link quality using FL [39], a selection of edge weights has been required. Therefore, in the fuzzy phase a combined Mamdani-Sugeno FL inference is used. When it comes to localization precision, this approach beats the traditional centroid.

The precision of these localization techniques is evaluated by ALE metrics. In order to properly tune the ALE below the appropriate threshold, we use an algorithm which produces minimal ALE value corresponding to the ideal network parameters. We have developed a powerful AI method for precise and very fast prediction of ALE in this situation to address this constraint. Basic localization of WSN nodes is shown in Figure 10.1.

Two alternative ANFIS models, namely the grid partition ANFIS model and the sub-clustering ANFIS model [40, 41] have been presented in this article. Four input features— anchor ratio, transmission range, node density, and number of iterations—are taken from the experimental dataset. In addition, this article is separated into six sections. We covered the system model for the node localization problem in Section 10.2. The ANFIS model and its implementation are described in Section 10.3. Result analysis is explained in Section 10.4 followed by conclusions in Section 10.5.

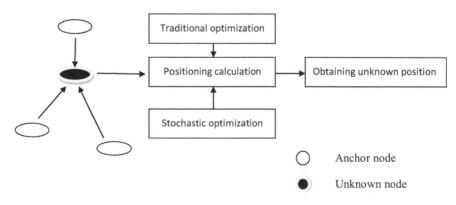

**FIGURE 10.1**
Localization of WSN nodes.

## 10.2 System Model

The system architecture created for the node localization procedure is discussed in this section. The process for calculating the distance between the anchor and unknown nodes is then covered. Later we discuss how the updated node localization algorithm's objective function was formed and how it operates. Finally, the two different ANFIS models designed to for this specific purpose are presented.

### 10.2.1 Distance Calculation for Generalization of Optimization Problem

The unknown nodes determine their separation from the anchor nodes using the RSSI. Due to multipath fading and shadowing [42], there is a power loss experienced by sensors during the information transmission. Path loss at a certain distance is a function of path loss at reference distance $d_0$, and the distance between TX and RX. In addition to this another component $\alpha$ also effect the path loss which indicates how the TX signal is fading due to increase the distance between TX and RX. The range $\alpha$ is set to 2 to 6 depends on a variety of physical factors, including antenna height, signal frequency, and propagation conditions [43]. There is a range error present at all times while determining the position of the unknown nodes. In order to evaluate the unknown nodes' positions as precisely as feasible, we take into account this inevitable range error.

The mean of the square of the inaccuracy between the actual distances of evaluated node coordinates and the estimated distance of actual unknown node coordinates from the nearby anchor nodes is used to calculate the optimization function (OF), which is the goal. There should be at least three anchor nodes within the transmission range of an unknown node in order to calculate its localization error. The evaluated position of the unknown node is the OF's lowest value.

### 10.2.2 Simulation Setup

In $100 \times 100$ m$^2$ square area, all the sensor nodes are distributed randomly. Different sensor node densities are implemented: 100, 200, and 300. The anchor nodes ratio is set to 14–30. The communication range for all sensor nodes is set to be 15–25 m and the number of iterations is set to 14–100. Before setting all these parameters, let's assume there is no localization error for any unknown node.

### 10.2.3 Experimental Results and Performance Analysis

#### 10.2.3.1 The Effect of Anchor Density

A key factor determining the effectiveness and cost of localization for WSN is anchor density [44]. This subsection assesses how anchor density affects the effectiveness of localization. Variously 10%, 20%, 30%, 40%, 50%, and 60% of all sensor nodes are designated as anchors. Each sensor node's communication range is 25 meters. From earlier research, it is clear that different anchor ratios (AR) in the network for various node densities are used to evaluate the ALE and confidence interval of location error (CILE). The anchor ratio can be increased from 10% to 40%, which will greatly enhance ALE [45, 46]. The number of unknown nodes that can realize within their communication range is increased by increasing the number of anchor nodes. The effects on the average localization error, however, become negligible while the anchor ratio keeps growing. In other words, there may not always be a need to add more expensive, specialized hardware-requiring anchor nodes.

### 10.2.3.2 The Effect of Communication Range

Communication range is another important parameter determining the localization error of an unknown node and energy consumption of sensor nodes [47]. 20% of the sensor nodes are set as the number of anchor nodes. In this research the variation of communication range is from 10 m to 25 m. According to earlier research, the communication range is just around 20 m, and the ALE is slightly larger as a result of the network's structure and lack of connectivity for many nodes. However, as more anchor information is supplied for determining the localization of unknown nodes, ALE is greatly reduced by expanding the communication range [49]. It is clear that as node density increases, the average localization error eventually decreases as well. This is because as node density rises, the number of anchor nodes that are accessible within communication range rises, as does the network connectivity between sensor nodes. Depending on the node density and communication range, the localization success ratio may vary [49, 50]. According to the experimental work, when the communication range is approximately 10 m and the node density is 100, a localization success ratio of roughly 20.8% may be reached. Because there are a substantial number of anchor pieces of information within the communication range, localization success ratio clearly improves as communication range grows. As a result, finding the unknown nodes is simpler. The localization requirements, such as localization accuracy, localization success ratio, and the energy constraint of sensor nodes, influence the choice of communication range.

## 10.3 Adaptive Neuro-Fuzzy Inference Architecture

ANFIS strategy is utilized to manage nondirect and complex problems [50, 51]. In an ANFIS hybrid intelligent system, a straightforward informational index produces desired output of a fuzzy logic controller through an interconnected neural network handling components by means of weighted data associations. ANFIS consolidates the quality of the two intelligent strategies FLC and NN into a solitary strategy [52, 53]. ANFIS model parameters of a FIS tuned by the neural network learning strategies. A five-layer ANFIS structure appears in Figure 10.2.

- It amends fuzzy *if-then* principles to depict the conduct of a nonstraight and complex framework.
- It doesn't require prior human skill.
- It is simple to actualize.
- It empowers quick and precise thinking and learning quality.
- Because of the legitimate choice of a reasonable decision of membership function, strong speculation, and brilliant clarification of fuzzy guidelines, it offers the desired output.
- It is simple to coordinate the both etymological and numeric information for critical thinking.

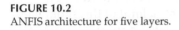

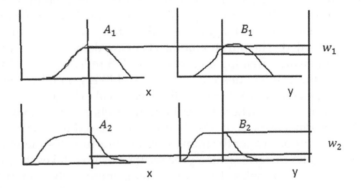

**FIGURE 10.2**
ANFIS architecture for five layers.

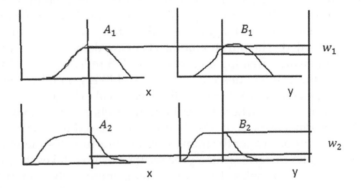

**FIGURE 10.3**
Firing of rules for different membership functions.

Figure 10.2 demonstrates the ANFIS engineering executed by the two principles where fixed nodes and versatile nodes are shown as circles and squares. Figure 10.3 represents a first order Sugeno model of ANFIS architecture. In Layer 1, every node is versatile, and the outputs of Layer 1 are the fuzzy membership of input. In Layer 2, the nodes are fixed and the fuzzy administrators fuzzify the inputs by utilizing AND operators. The symbol Π shows a basic multiplier activity. Output of layer 2 provides the standardized firing qualities. In Layer 3, fixed nodes named as N normalize the firing strengths from the previous layer. In Layer 4, nodes are versatile, and the output of every node is basically a standardized firing quality and a first order polynomial. Layer 5 only consists of a single fixed node Σ which plays out the summation of every single approaching sign.

### 10.3.1 Hybrid Learning ANFIS

In this algorithm the ANFIS model is a blend of least squares strategies and gradient descent strategies. In the forward pass learning calculation, least square techniques are utilized to determine node outputs until Layer 4 and the consequent parameters. In the backward path, the gradient descendent method sends the error signals to the previous stage and updates parameters. The hybrid learning approach is quicker than the original back-propagation technique because of diminishing search space dimensions. The ideal estimation of these parameters is controlled by the least squares technique. At the point when these parameters are variable, the search space in hybrid learning process increases; subsequently, the convergence rate of the training datasets becomes slower. To tackle the issue of search space, the hybrid algorithm of ANFIS joins two techniques: (1) least squares strategy and (2) gradient descent technique, where the least squares strategy is used to advance the subsequent parameters and the gradient descent technique is utilized to streamline the reason parameters. From the review it has been seen that the hybrid algorithm gives a high level of proficiency in preparing the ANFIS frameworks.

### 10.3.2 ANFIS Training Process

The ANFIS strategy starts by taking the two distinctive datasets, training and testing, where the training dataset comprises info and yield vectors. A membership function of the fuzzy model is implemented by the training dataset and acquires a limit of threshold value with the help of examination between experimental and calculated output by applying a least square technique; however, in the event that the error between the experimental and calculated value is huge, at that point the gradient decent method automatically refreshes the membership function until the magnitude of the error is just less than threshold value and the procedure is ended. The purpose of the checking dataset is to contrast the model and the genuine framework. Training datasets of ANFIS re learned by a mix of the least squares strategy and the gradient descent strategy. ANFIS is generally utilized in a versatile process control framework to accomplish the most ideal exhibition. Figure 10.4 depicts the working process of ANFIS, where the training datasets are used to model the ANFIS in the MATLAB platform. Datasets of ANFIS model are placed in a matrix, where the last column indicates the output. As per the framework architecture, the membership function is created. By utilizing the correct learning process, the designer takes the proper membership function of the input variables. The initial arrangement of the membership function works likewise and is made by utilizing the direction genfis in MATLAB [53]. The framework begins to training after the underlying membership function is made. In the wake of utilizing the fismat direction, input information is prepared and the membership function naturally builds. After termination of the training process, the final membership functions of the input variable and training error are produced. To build the exactness of the model, checking datasets can be utilized. The ANFIS model can be utilized for just one training dataset; however, the framework's viability can be expanded by applying the checking datasets in a framework.

In the present research, the evalfis work in ANFIS is utilized to ponder and assess the framework execution of average localization error (ALE) in WSN. At first in this model input datasets are used to design the fuzzy framework, in spite of the fact that these input datasets do not exclude any output esteems. The output of the evalfis capacity gives the final output of the ANFIS system. The relationships between the experimental and calculated ALE are set after the preparation estimated the system output. When the model is prepared, we can further test the framework against various arrangements of input datasets to check the usefulness. The training steps are clarified in Figure 10.4, where the forecast of ALE shows the ANFIS model assisted by four input process variables and one output variable [54–55]. Learning rules of the ANFIS model thoroughly depend upon the input and output variable membership function, set by the human expert. The anchor ratio (AR), transmission range (TR), node density (ND), and number of iterations (IT) have three

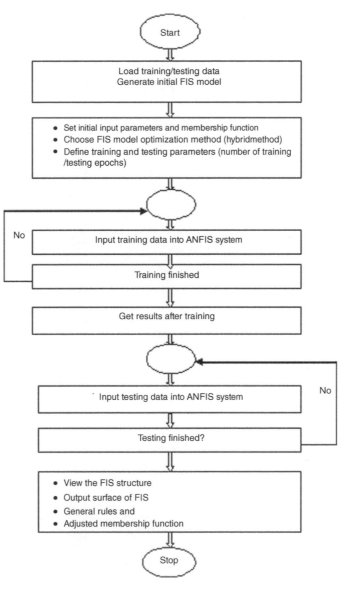

**FIGURE 10.4**
Flowchart of ANFIS training system.

membership functions named as high, medium, and low, while for output parameters, the ALE has three membership functions, the same as for input variables. The ANFIS preparing procedure begins by the fuzzy sets, number of information factors, and state of the enrollment capacity of the information factors.

Here, N and $e_i$ are the total number of predictions and the difference between predicted and original series output, respectively. In the ANFIS structure, four information parameters and a single output controlled parameter, liquid flow rate, appear in Figure 10.5. Error in training informational indexes and testing datasets are shown in Figures 10.6 and 10.7, respectively. From the error diagram it is seen that preparation error and testing error both are decreased by expanding the number of epochs. Figure 10.7 demonstrates the standard

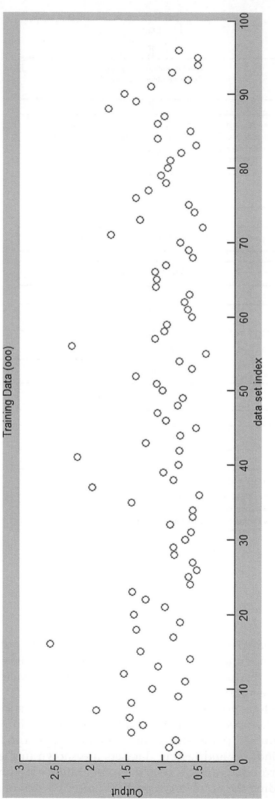

**FIGURE 10.5**
Distribution of training dataset with respect to dataset index.

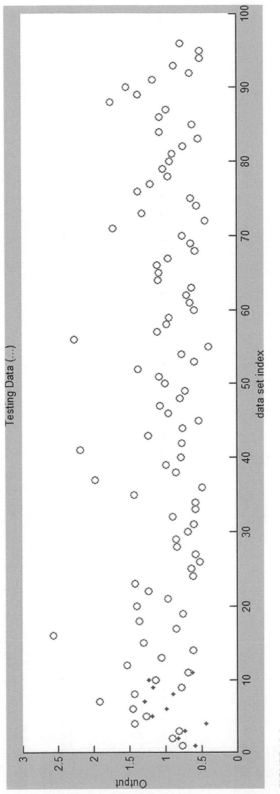

**FIGURE 10.6**
Distribution of testing dataset with respect to dataset index.

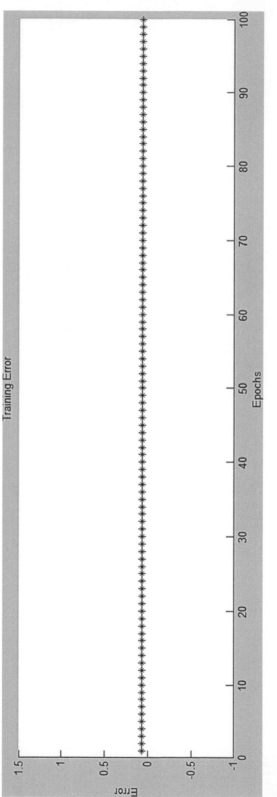

**FIGURE 10.7**
Training error with respect to number of epochs (training error 0.0693 after 100 epochs).

rule viewers of the present fluid stream control framework between the four information factors and one yield controlled variable in ANFIS model.

## 10.4 Result Analysis

Overall, research authors have used two different types of ANFIS model: grid partition methods and subclustering methods.

### 10.4.1 Grid Partition Method

Before implement of the ANFIS model, the overall 107 member dataset was partitioned into two sections: a training dataset that contained 96 items and a testing or target dataset containing 11 items. Training and testing data characteristics are presented with respect to the data index in Figure 10.5 and 10.6, respectively. Each of the input parameters of unknown sensor variables has three membership functions; hence there should be 81 possible combinations between process input and output variables as shown in Figure 10.8. In the rule viewer for each combination of input variables, the output variable, namely average location error, can be achieved, which is shown in Figure 10.9. Figures 10.10–10.12 show how the unknown sensor output variable ALE depends upon the possible combination of input variables such as anchor ratio (AR), transmission range (TR), node density (ND), and number of iterations (IT).

```
1. If (input1 is in1mf1) and (input2 is in2mf1) and (input3 is in3mf1) and (input4 is in4mf1) then (output is out1mf1) (1)
2. If (input1 is in1mf1) and (input2 is in2mf1) and (input3 is in3mf1) and (input4 is in4mf2) then (output is out1mf2) (1)
3. If (input1 is in1mf1) and (input2 is in2mf1) and (input3 is in3mf1) and (input4 is in4mf3) then (output is out1mf3) (1)
4. If (input1 is in1mf1) and (input2 is in2mf1) and (input3 is in3mf2) and (input4 is in4mf1) then (output is out1mf4) (1)
5. If (input1 is in1mf1) and (input2 is in2mf1) and (input3 is in3mf2) and (input4 is in4mf2) then (output is out1mf5) (1)
6. If (input1 is in1mf1) and (input2 is in2mf1) and (input3 is in3mf2) and (input4 is in4mf3) then (output is out1mf6) (1)
7. If (input1 is in1mf1) and (input2 is in2mf1) and (input3 is in3mf3) and (input4 is in4mf1) then (output is out1mf7) (1)
8. If (input1 is in1mf1) and (input2 is in2mf1) and (input3 is in3mf3) and (input4 is in4mf2) then (output is out1mf8) (1)
9. If (input1 is in1mf1) and (input2 is in2mf1) and (input3 is in3mf3) and (input4 is in4mf3) then (output is out1mf9) (1)
10. If (input1 is in1mf1) and (input2 is in2mf2) and (input3 is in3mf1) and (input4 is in4mf1) then (output is out1mf10) (1)
11. If (input1 is in1mf1) and (input2 is in2mf2) and (input3 is in3mf1) and (input4 is in4mf2) then (output is out1mf11) (1)
12. If (input1 is in1mf1) and (input2 is in2mf2) and (input3 is in3mf1) and (input4 is in4mf3) then (output is out1mf12) (1)
13. If (input1 is in1mf1) and (input2 is in2mf2) and (input3 is in3mf2) and (input4 is in4mf1) then (output is out1mf13) (1)
14. If (input1 is in1mf1) and (input2 is in2mf2) and (input3 is in3mf2) and (input4 is in4mf2) then (output is out1mf14) (1)
15. If (input1 is in1mf1) and (input2 is in2mf2) and (input3 is in3mf2) and (input4 is in4mf3) then (output is out1mf15) (1)
16. If (input1 is in1mf1) and (input2 is in2mf2) and (input3 is in3mf3) and (input4 is in4mf1) then (output is out1mf16) (1)
17. If (input1 is in1mf1) and (input2 is in2mf2) and (input3 is in3mf3) and (input4 is in4mf2) then (output is out1mf17) (1)
18. If (input1 is in1mf1) and (input2 is in2mf2) and (input3 is in3mf3) and (input4 is in4mf3) then (output is out1mf18) (1)
```

If                                    and                                    and

**FIGURE 10.8**
Rule editor for the input and output system variables containing 81 rules.

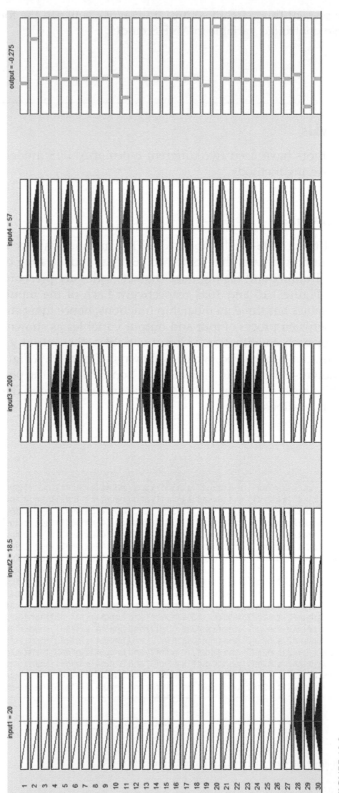

**FIGURE 10.9**
Rule viewers for the input and output system process variables.

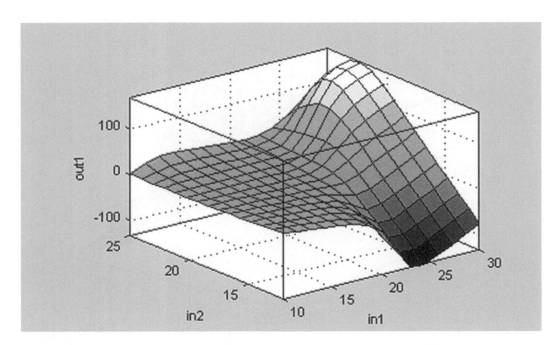

**FIGURE 10.10**
3D surface view of anchor ratio (AR), transmission range (TR), and ALE.

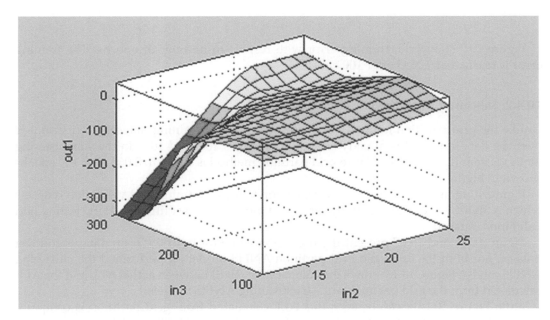

**FIGURE 10.11**
3D surface view of transmission range (TR), node density, and ALE.

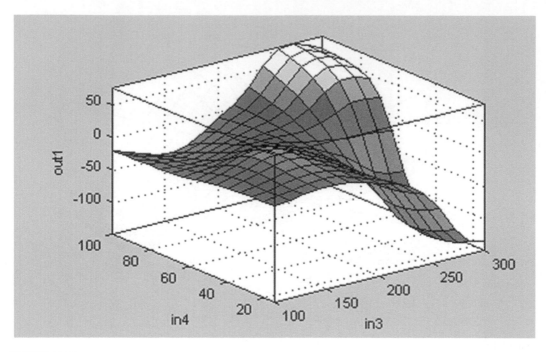

**FIGURE 10.12**
3D surface views of node density, iteration, and ALE.

Figure 10.7 shows the training error with respect to number of epochs. The training error is reached at 0.0693 after 100 epochs.

### 10.4.2 Subclustering Method

Unlike the grid partition method, all the test dataset and training dataset are distributed over the data index value already as shown in Figure 10.5 and 10.6. In the subclustering method, training error is similar to grid partition method is about to 0.069 after 100 epochs shown in Figure 10.13.

Figure 10.14 and 10.15 present the regression value of training and validation datasets which is about 1, indicating both the proposed models are best fitted with the testing and validation dataset.

Figure 10.16 and 10.17 present the regression value for both the train dataset and test dataset. For both the cases the grid partition ANFIS model outperformed the clustering ANFIS model. Figure 10.18 shows the comparative result between actual ALE and the ALE calculated from the grid partition and subclustering ANFIS models.

From Figure 10.18 it can be seen that ALE calculated from grid partition is comparatively better fitted than the ALE calculated from the cluster ANFIS model.

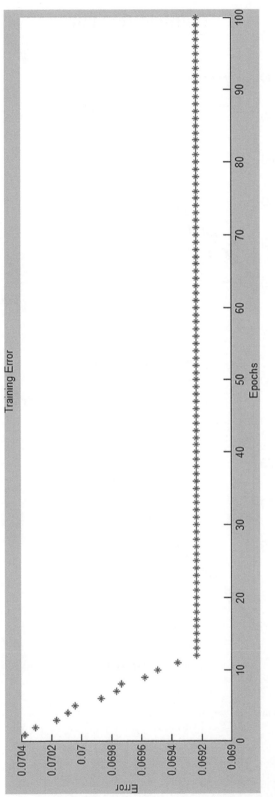

**FIGURE 10.13**
Training error 0.069 after 100 epochs.

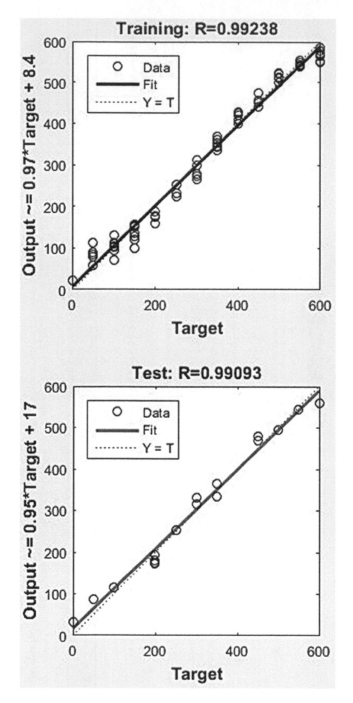

**FIGURE 10.14**

Regression value for after training data set.

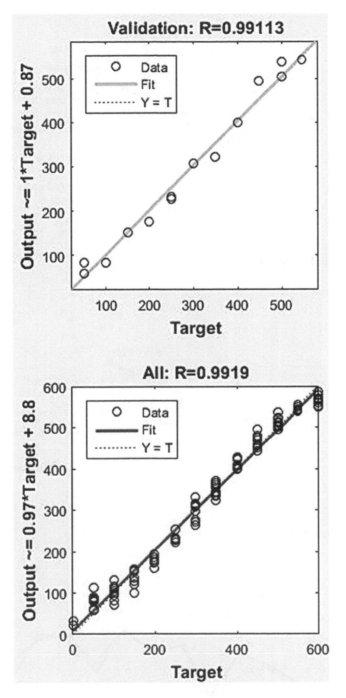

**FIGURE 10.15**
Regression value after validation.

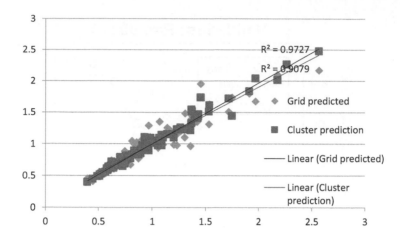

**FIGURE 10.16**
Regression model for the training dataset.

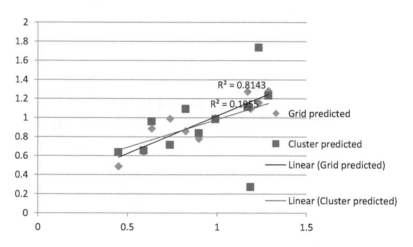

**FIGURE 10.17**
Regression model for Test dataset.

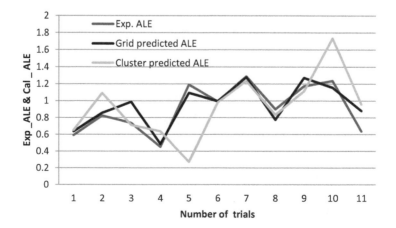

**FIGURE 10.18**
Comparative result for experimental ALE, grid partition ALE, and cluster ALE.

## 10.5 Conclusions

Demonstration and advancement of advanced wireless communication networks is an intriguing assignment for researchers. In wireless sensor networking systems, implementation of any unknown is very important for proper power transmission between TX and RX. To get the optimum localization error of unknown nodes depends upon the four input potential parameters, namely transmission range (TR), node density (ND), anchor ratio (AR), and iterations (IT). In this research to get the optimum average localization error, 107 datasets were used to implement the model, and for the prediction of the model, a new ANFIS model is used. Overall result analysis is carried out by the two different ANFIS models, namely grid partition method and subclustering method.

From the result analysis, it is concluded that both the algorithms achieved the same magnitude of training error, but for the forecasting of ALE of unknown sensor nodes in the testing dataset, the grid partition based ANFIS model was best fitted compared to the subclustering method.

The following inferences are drawn from the graphs:

- The type and quantity of membership functions are crucial when developing an ANFIS architecture.
- As a system output, the number of membership functions and training data samples has a favorable impact on the ALE.
- Increasing the number of membership functions results in over fitting but has no effect on the model's performance.
- A larger training sample size results in more respectable outcomes.
- The number of epochs aids in preventing over fitting.
- The test's results show what is necessary for improved performance.

Besides the ANFIS calculation, how the metaheuristics streamlining strategy is utilized to improve the productivity, exactness, convergence speed, security, and achievement pace of the present procedure control is an additional future research direction.

## References

1. Kandris D, Nakas C, Vomvas D, Koulouras G. Applications of wireless sensor networks: An up-to-date survey. *Applied System Innovation* 2020;3(1): 14.
2. Khalaf OI, Sabbar BM. An overview on wireless sensor networks and finding optimal location of nodes. *Periodicals of Engineering and Natural Sciences (PEN)* 2019;7(3): 1096–1101.
3. Sah DK, Amgoth T. Renewable energy harvesting schemes in wireless sensor networks: A survey. *Information Fusion* 2020;63: 223–247.
4. Adu-Manu KS, Adam N, Tapparello C, Ayatollahi H, Heinzelman W. Energy-harvesting wireless sensor networks (EH-WSNs): A review. *ACM Transactions on Sensor Networks (TOSN)* 2018;14(2): 1–50.
5. Rajasekaran T, Anandamurugan S. Challenges and applications of wireless sensor networks in smart farming—A survey. In *Advances in big data and cloud computing.* Springer;2019, pp. 353–361.

6. Abdulkarem M, Samsudin K, Rokhani FZ, Rasid AMF. Wireless sensor network for structural health monitoring: A contemporary review of technologies, challenges, and future direction. *Structural Health Monitoring* 2020;19(3): 693–735.

7. Delavernhe F, Lersteau C, Rossi A, Sevaux M. Robust scheduling for target tracking using wireless sensor networks. *Computers & Operations Research* 2020;116: 104873.

8. Ullo S, Gallo M, Palmieri G, Amenta P, Russo M, Romano G, … De Angelis M. Application of wireless sensor networks to environmental monitoring for sustainable mobility. In *2018 IEEE International Conference on Environmental Engineering (EE)*. IEEE;2018, March, pp. 1–7.

9. Khujamatov KE, Toshtemirov TK. Wireless sensor networks based agriculture 4.0: Challenges and apportions. In *2020 International Conference on Information Science and Communications Technologies (ICISCT)*.IEEE;2020, November, pp. 1–5.

10. Muduli L, Mishra DP, Jana PK. Application of wireless sensor network for environmental monitoring in underground coal mines: A systematic review. *Journal of Network and Computer Applications* 2018;106: 48–67.

11. Qasim HH, Hamza AE, Ibrahim HH, Saeed HA, Hamzah MI. Design and implementation home security system and monitoring by using wireless sensor networks WSN/internet of things IOT. *International Journal of Electrical and Computer Engineering* 2020;10(3): 2617.

12. Kang B, Choo H. An energy efficient routing scheme by using GPS information for wireless sensor networks. *International Journal of Sensor Networks* 2018;26: 136–143.

13. Khalaf OI, Sabbar BM. An overview on wireless sensor networks and finding optimal location of nodes. *Periodicals of Engineering and Natural Sciences (PEN)* 2019;7(3): 1096–1101.

14. Sneha V, Nagarajan M. Localization in wireless sensor networks: A review. *Cybernetics and Information Technologies* 2020;20(4): 3–26.

15. Singh A, Kotiyal V, Sharma S, Nagar J, Lee CC. A machine learning approach to predict the average localization error with applications to wireless sensor networks. *IEEE Access* 2020;8: 208253–208263.

16. Ismail MIM, Dzyauddin RA, Samsul S, Azmi NA, Yamada Y, Yakub MFM, Salleh NABA. An RSSI-based wireless sensor node localisation using trilateration and multilateration methods for outdoor environment. 2019. arXiv:1912.07801.

17. Zaidi M, Bouazzi I, Al-Rayif MI, Shamim MZM, Usman M. Low power hardware design and its mathematical modeling for fast-exact geolocalization system in wireless networks. *International Journal of Communication Systems* 2022;35(9): e5128.

18. Sivasakthiselvan S, Nagarajan V. Localization techniques of wireless sensor networks: A review. In *2020 International Conference on Communication and Signal Processing (ICCSP)*. IEEE;2020, July, pp. 1643–1648.

19. Shahra EQ, Sheltami TR, Shakshuki EM. A comparative study of range-free and range-based localization protocols for wireless sensor network: Using cooja simulator. In *Sensor technology: Concepts, methodologies, tools, and applications*. IGI Global;2020, pp. 1522–1537.

20. Silmi S, Doukha Z, Moussaoui S. A self-localization range free protocol for wireless sensor networks. *Peer-to-Peer Networking and Applications* 2021;14(4): 2061–2071.

21. Gui L, Huang X, Xiao F, Zhang Y, Shu F, Wei J, Val T. DV-Hop localization with protocol sequence based access. *IEEE Transactions on Vehicular Technology* 2018;67(10): 9972–9982.

22. Silmi S, Doukha Z, Moussaoui S. A self-localization range free protocol for wireless sensor networks. *Peer-to-Peer Networking and Applications* 2021;14(4): 2061–2071.

23. Saad E, Elhosseini M, Haikal AY. Recent achievements in sensor localization algorithms. *Alexandria Engineering Journal* 2018;57(4): 4219–4228.

24. Al-Madani BM, Shahra EQ. An energy aware plateform for IoT indoor tracking based on RTPS. *Procedia Computer Science* 2018;130: 188–195.

25. Akram J, Javed A, Khan S, Akram A, Munawar HS, Ahmad W. Swarm intelligence based localization in wireless sensor networks. In *Proceedings of the 36th Annual ACM Symposium on Applied Computing*. 2021, March, 1906–1914.

26. Malhotra S, Matta P, Kukreti S. Efficient energy consumption techniques for cloud and IoT systems. In *2022 IEEE Delhi Section Conference (DELCON)*. IEEE;2022, February, pp. 1–7.

27. Dutta P, Majumder M, Kumar A. Parametric optimization of liquid flow process by ANOVA optimized DE, PSO & GA algorithms. *International Journal of Image Graphics and Signal Processing* 2021;11: 14–24.

28. Dutta P, Majumder M, Kumar A. An improved grey wolf optimization algorithm for liquid flow control system. *International Journal of Engineering and Manufacturing* 2021;4: 10–21.

29. Dutta P, Kumar A. Modelling of liquid flow control system using optimized genetic algorithm. *Statistics, Optimization & Information Computing* 2020;8(2): 565–582.

30. Dutta P, Kumar A. Modeling and optimization of a liquid flow process using an artificial neural network-based flower pollination algorithm. *Journal of Intelligent Systems* 2020;29(1): 787–798.

31. Dutta P, Paul S, Kumar A. Comparative analysis of various supervised machine learning techniques for diagnosis of COVID-19. In *Electronic devices, circuits, and systems for biomedical applications*. Academic Press;2021, pp. 521–540.

32. Dutta P, Biswas SK, Biswas S, Majumder M. Parametric optimization of solar parabolic collector using metaheuristic optimization. *Computational Intelligence and Machine Learning* 2021;2: 26–32.

33. Dutta P, Paul S, Obaid AJ, Pal S, Mukhopadhyay K. Feature selection based artificial intelligence techniques for the prediction of COVID like diseases. In *Journal of Physics: Conference Series*. IOP Publishing;2021, July, Vol. 1963, No. 1, p. 012167.

34. Dutta P, Kumar A. Application of an ANFIS model to optimize the liquid flow rate of a process control system. *Chemical Engineering Transactions* 2018;71: 991–996.

35. Dutta P, Agarwala R, Majumder M, Kumar A. Parameters extraction of a single diode solar cell model using bat algorithm, firefly algorithm & cuckoo search optimization. *Annals of the Faculty of Engineering Hunedoara-International Journal of Engineering* 2020;18: 147–156.

36. Janssen T, Berkvens R, Weyn M. Benchmarking RSS-based localization algorithms with LoRaWAN. *Internet of Things* 2020;11: 100235.

37. Aldosari W, Zohdy M, Olawoyin R. Tracking the mobile jammer in wireless sensor networks using extended Kalman filter. In *2019 IEEE 10th Annual Ubiquitous Computing, Electronics & Mobile Communication Conference (UEMCON)*. IEEE;2019, October, pp. 0207–0212.

38. Hadir A, Zine-Dine K, Bakhouya M. Improvements of Centroid Localization Algorithm for Wireless Sensor Networks. In *2020 5th International Conference on Cloud Computing and Artificial Intelligence: Technologies and Applications (CloudTech)*. IEEE;2020, November, pp. 1–6.

39. Dutta P, Kumar A. Design an intelligent flow measurement technique by optimized fuzzy logic controller. *Journal Europen des Systmes Automatiss* 2018;51: 89–107.

40. Adeleke O, Akinlabi SA, Jen TC, Dunmade I. Prediction of municipal solid waste generation: An investigation of the effect of clustering techniques and parameters on ANFIS model performance. *Environmental Technology* 2022;43(11): 1634–1647.

41. Gill J, Singh J, Ohunakin OS, Adelekan DS, Atiba OE, Nkiko MO, Atayero AA. Adaptive neuro-fuzzy inference system (ANFIS) approach for the irreversibility analysis of a domestic refrigerator system using LPG/TiO 2 nanolubricant. *Energy Reports* 2020;6: 1405–1417.

42. Nguyen NM, Tran LC, Safaei F, Phung SL, Vial P, Huynh N, … Barthelemy J. Performance evaluation of non-GPS based localization techniques under shadowing effects. *Sensors* 2019;19(11): 2633.

43. Messous S, Liouane H, Cheikhrouhou O, Hamam H. Improved recursive DV-hop localization algorithm with RSSI measurement for wireless sensor networks. *Sensors* 2021;21(12): 4152.

44. Cheng J, Li Y, Xu Q. An anchor node selection scheme for improving RSS-based localization in wireless sensor network. *Mobile Information Systems* 2022;2022: 2611329.

45. Liu X, Han F, Ji W, Liu Y, Xie Y. A novel range-free localization scheme based on anchor pairs condition decision in wireless sensor networks. *IEEE Transactions on Communications* 2020;68(12): 7882–7895.

46. Hamouda E, Abohamama AS. Wireless sensor nodes localiser based on sine–cosine algorithm. *IET Wireless Sensor Systems* 2020;10(4): 145–153.

47. Han D, Yu Y, Li KC, de Mello RF. Enhancing the sensor node localization algorithm based on improved DV-hop and DE algorithms in wireless sensor networks. *Sensors* 2020;20(2): 343.

48. Bhat SJ, Venkata SK. An optimization based localization with area minimization for heterogeneous wireless sensor networks in anisotropic fields. *Computer Networks* 2020;179: 107371.

49. Phoemphon S, So-In C, Leelathakul N. Fuzzy weighted centroid localization with virtual node approximation in wireless sensor networks. *IEEE Internet of Things Journal* 2018;5(6): 4728–4752.

50. AlRassas AM, Al-qaness MA, Ewees AA, Ren S, Abd Elaziz M, Damaševičius R, Krilavičius T. Optimized ANFIS model using Aquila optimizer for oil production forecasting. *Processes* 2021;9(7): 1194.

51. Armaghani DJ, Asteris PG. A comparative study of ANN and ANFIS models for the prediction of cement-based mortar materials compressive strength. *Neural Computing and Applications* 2021;33(9): 4501–4532.

52. Sada SO, Ikpeseni SC. Evaluation of ANN and ANFIS modeling ability in the prediction of AISI 1050 steel machining performance. *Heliyon* 2021;7(2): e06136.

53. Zhu H, Zhu L, Sun Z, Khan A. Machine learning based simulation of an anti-cancer drug (busulfan) solubility in supercritical carbon dioxide: ANFIS model and experimental validation. *Journal of Molecular Liquids* 2021;338: 116731.

54. Aghbashlo M, Tabatabaei M, Nadian MH, Davoodnia V, Soltanian S. Prognostication of lignocellulosic biomass pyrolysis behavior using ANFIS model tuned by PSO algorithm. *Fuel* 2019;253: 189–198.

55. Sremac S, Tanackov I, Kopić M, Radović D. ANFIS model for determining the economic order quantity. *Decision Making: Applications in Management and Engineering* 2018;1(2): 81–92.

# 11

## Performance Estimation of Photovoltaic Cell Using Hybrid Genetic Algorithm and Particle Swarm Optimization

### 11.1 Introduction

The energy shortage and the adverse consequences of pollution make renewable energy like solar energy more attractive in the modern era [1–3]. Simulation of the solar cell is crucial to estimate the performance given the attributes of current–voltage and power–voltage in a solar cell under various solar intensities and temperature conditions. There are several studies that have been performed regarding the solar cell, but the most common literature survey is performed around the double diode model (DDM) and single diode model (SDM) [4–8]. In a photovoltaic (PV) system, estimation of the optimal parameter is required for the prognosis of a solar cell's efficacy; for that, the optimal optimization approach is needed [9, 10].

One of the foremost prominent subclasses of optimization algorithms is metaheuristics, where simplifying patterns are commonly spurred by scientific observations, creature behaviors, or evolutionary assumptions. The topic of parameter evaluation of PV cells or modules has been addressed using a multitude of metaheuristic approaches or their adaptations, including genetic programming [11, 12], differential evolutionary [13, 14], ant bee colony (ABC) [15, 16], chaotic Jaya algorithm (JAYA) [17], teaching learning based optimization (TLBO) [18, 19], shuffled frog leaping algorithm (SFLA) [20], moth–flame optimization calculation [21], ant lion optimization [22], sine–cosine algorithm [23], grey wolf optimization (GWO) [24], flower pollination algorithm (FPA) [25, 26], improved elephant swarm water search calculation [27, 28], particle swarm optimization (PSO) [29], and so on. A comparative study is performed for parametric evacuation of a single solar cell utilizing three metaheuristic optimizations; the bat algorithm, cuckoo optimization, and firefly algorithm [30]. Parametric optimization for a double diode solar cell is performed by wind-driven optimization [31]. For the identified issue, these metaheuristic assessments have been accomplished magnificently. In addition to these metaheuristic optimization techniques, other AI techniques like fuzzy logic controller [32, 33], ANN model [34], genetic algorithm (GA) [35], and so on can be offered in enhancing the variables of SDM and DDM models.

Whatever the case, accepting the no free lunch principle [36], there are no efficient single metaheuristics for addressing a plethora of difficulties. This is why it is still crucial for the researcher to quantify the specifications of PV cells or panels. The improved whale optimization algorithm (IWOA) [39], biogeography-based heterogeneous cuckoo search

DOI: 10.1201/9781003216001-13

algorithm [40], self-adaptive ensemble-based differential evolution [41], hybrid differential evolution with whale optimization algorithm [42], improved chaotic whale optimization algorithm [43], and hybrid bee pollinator flower pollinator were utilized for process parameters of a double diode, single diode, and PV model [45], along with hybrid firefly and pattern search algorithms [46] and hybrid PSO and GWO algorithm [47, 48]. To get the optimum parameters, convergence speed, and computational time, other improved optimization techniques can be applied, and this is still an open challenge for present research.

Particle swarm optimization (PSO) is an uncomplicated technique, simple to carry out, computationally adequate, restricted in the local minimum, and dealing with limited local/global search abilities in certain constraints [49–51]. However, its solution may be trapped in local optima, and due to fast convergence, its results may not be accurate [52]. On the other hand, the genetic algorithm provides the global optimum solution, but it requires a significant amount of computational time as it provides the better result by increasing the number of iteration as well as the number of computational steps [35, 53–55]. Consequently, we present an integrated optimization research methodology by taking the advantages of both PSO and GA.

The hybrid GA-PSO (HGAPSO) method is predicted to outperform other algorithms with similar objectives to function faster with varied sizes of workflow applications. Furthermore, because the GA mutation operator is used to improve the accuracy of the solutions utilized to identify the answer for many complicated and nonlinear problems, the hybrid GA-PSO algorithm may not get caught in the local optimal solution.

The rest of this chapter is organized as follows. The scientific representation of the solar cell is illustrated in Section 11.2, and the synthesis of the goal functioning of the double and single diode models is demonstrated in Section 11.3. The proposed mixture of HGAPSO is expounded in Section 11.4. Next, re-enacted results and discussion of experimental results with the key study are given in Section 11.5. Finally, Section 11.6 closes this paper, followed by the references.

## 11.2 Mathematics Model and Objective Function of the Solar Cell

### 11.2.1 Single Diode Model (SDM)

The equivalent circuit diagram of SDM [48] is portrayed in Figure 11.1. Output current of SDM solar cell can be figured as follows:

$$I_C = I_{ph} - I_d - I_{sh} \tag{11.1}$$

Where $I_C$, $I_{ph}$, $I_d$, and $I_{sh}$ are the cell output current, photogenerated current, diode current, and shunt resistor current, respectively.

According to the Shockley equation, $I_d$ can be calculated as

$$I_d = I_{sd}\left[\exp\left(\frac{q(V_C + I_C R_s)}{\eta KT}\right) - 1\right] \tag{11.2}$$

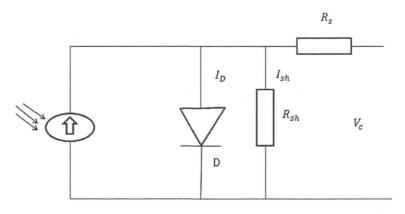

**FIGURE 11.1**
Single diode model [48].

Where $I_{sd}$, $V_C$, $\eta$ and $R_s$ are the reverse saturation current, cell output voltage, ideality factor, and series resistance, respectively.

Let $V_t = KT/q$; $K$ is the Boltzmann constant, $T$ is the absolute temperature (Kelvin), and $q$ is the electron charge. Then Eq. (11.2) can be further modified into

$$I_d = I_{sd}\left[\exp\left(\frac{V_C + I_C R_s}{\eta V_t}\right) - 1\right] \tag{11.3}$$

The current passing through the shunt resistor $I_{sh}$ is formulated as

$$I_{sh} = \frac{V_C + I_C R_s}{R_{sh}} \tag{11.4}$$

After combining Eqs. (11.1), (11.3) and (11.4), the I-V relationship of the SDM can be communicated as

$$I_C = I_{ph} - I_{sd}\left[\exp\left(\frac{V_C + I_C R_s}{\eta V_t}\right) - 1\right] - \frac{V_C + I_C R_s}{R_{sh}} \tag{11.5}$$

Hence from Equation 11.5, it can be seen that it contains five parameters ($I_{ph}$, $I_{sd}$, $R_s$, $R_{sh}$, $\eta$) which ought to be estimated by the optimization tool.

### 11.2.2 Double Diode Model (DDM)

The equivalent circuit diagram of DDM [48] is portrayed in Figure 11.2. Cell output current $I_C$ can be representing by the following equation:

$$I_C = I_{ph} - I_{d1} - I_{d2} - I_{sh} \tag{11.6}$$

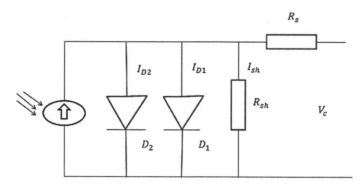

**FIGURE 11.2**
Double diode model [48].

In DDM, the V-I relationship can be finally communicated as

$$I_C = I_{\text{ph}} - I_{\text{sd1}}\left[\exp\left(\frac{V_C + I_C R_s}{\eta_1 V_t}\right) - 1\right] - I_{\text{sd2}}\left[\exp\left(\frac{V_C + I_C R_s}{\eta_2 V_t}\right) - 1\right] - \frac{V_C + I_C R_s}{R_{\text{sh}}} \qquad (11.7)$$

From Eq. (11.7) it is seen that the model contains seven parameters which have to be estimated ($I_{\text{ph}}, I_{\text{sd1}}, I_{\text{sd2}}, R_s, R_{\text{sh}}, \eta_1, \eta_2$).

The reduced objective function relates to better assessed parameters. The nonlinear and transcendental objective function is hard to settle using the conventional method.

### 11.2.3 PV Module Model

A PV module's double-diode and single-diode variants, which is comprised of linked cells in series, can also be articulated as Eqs. (11.5) and (11.7), where

$$V_t = N_s KT / q.$$

## 11.3 Objective Function

The estimated current–voltage pattern of a PV system is often fitted to the observed pattern using metaheuristics. The goal of the assessment is to determine the critical parameters' appropriate options in order to reduce the difference in current between the measured and simulated current. Equation (11.8) defines the goal function as the root mean square of the error (RMSE) [56].

$$\text{RMSE}(X) = \sqrt{\frac{1}{N}\sum_{i=1}^{N} f\left(V_{\text{Ci}}, I_{\text{Ci}}, X\right)^2} \qquad (11.8)$$

Where $N$ stands for the quantity of scientific results, and $X$ stands for the set of predictor variables.

For the single diode model, $f(V_C, I_C, X)$ and $X$ can be respectively expressed as Eqs. (11.9) and (11.10).

$$f(V_C, I_C, X) = I_{ph} - I_{sd}\left[\exp\left(\frac{V_C + I_C R_s}{\eta V_t}\right) - 1\right] - \frac{V_C + I_C R_s}{R_{sh}} - I_C \tag{11.9}$$

$$\text{RMSE}_{min} = \sqrt{\frac{1}{N}\sum_{i=1}^{N}\left(I_{measured} - I_{calculated}\left(I_{ph}, I_{sd}, R_s, R_{sh}, \eta\right)\right)^2} \tag{11.10}$$

$$X = \{I_{ph}, I_{sd}, R_s, R_{sh}, \eta\}$$

For the double diode model, $f(V_C, I_C, X)$ and $X$ can be correspondingly exhibited as Eqs. (11.11) and (11.12).

$$f(V_C, I_C, X) = I_{ph} - I_{sd1}\left[\exp\left(\frac{V_C + I_C R_s}{\eta_1 V_t}\right) - 1\right] - I_{sd2}\left[\exp\left(\frac{V_C + I_C R_s}{\eta_2 V_t}\right) - 1\right] - \frac{V_C + I_C R_s}{R_{sh}} - I_C \tag{11.11}$$

$$\text{RMSE}_{min} = \sqrt{\frac{1}{N}\sum_{i=1}^{N}\left(I_{measured} - I_{calculated}\left(I_{ph}, I_{sd1}, I_{sd2}, R_s, R_{sh}, \eta_1, \eta_2\right)\right)^2} \tag{11.12}$$

$$X = \{I_{ph}, I_{sd1}, I_{sd2}, R_s, R_{sh}, \eta_1, \eta_2\}$$

The optimum value that is lower has parameters that are more accurately predicted. This task is challenging to tackle since the goal function is nonlinear and transcendent.

## 11.4 Proposed Methodology

### 11.4.1 Improved Cuckoo Search Optimization

This section introduces a hybrid algorithm that merges the GA and PSO strategies. The schematic in Figure 11.3 provides the indispensable steps of the HGAPSO algorithm. PSO are among the most effective methodologies; however, because of how efficiently it converges, it typically adheres erroneously in convoluted scenarios. The GA algorithm surpassed other strategies when employed to tackle a variety of intricate difficulties. Although the GA method may settle somewhat more slowly, it has higher probing flexibility. No strategy is adequate in treating all scalability issue effectively. The problem may be solved by integrating the existing approach to achieve the aggregate finest resolution [57, 58]. The proposed technique can boost global retrieval accuracy and avoid premature convergence. It could make it less likely to become locked in a local optimal solution. The best elements of both technologies' characteristics may be combined in the hybrid algorithm (Figure 11.3).

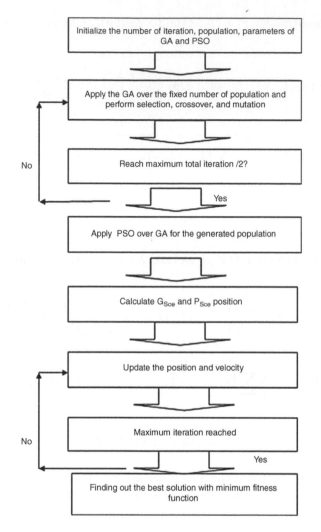

**FIGURE 11.3**
Flowcharts for the proposed HGAPSO algorithm.

## 11.5 Results and Discussion

### 11.5.1 Test Information

In this section, we have taken test datasets of double diode and single diode RTC France silicon solar cells with 26 sets of I-V data at a temperature of 33°C and irradiation of 1000 W/m$^2$ [30, 48]. The effects of the HGAPSO optimization approach on the two previously described benchmark difficulties are presented in this part. Additionally, we compare HGAPSO to the other two fundamental methodologies, PSO and GA, and provide a rational assessment of the evaluated results. PSO and GA are two well-known optimization strategies that we have adopted for reference purposes; the specifications made for each algorithm are described as follows (Table 11.1):

**TABLE 11.1**

Parameter Setting for Each Algorithm

| PSO | GA | HGAPSO |
|---|---|---|
| Number of population $N = 100$, inertia weight may be between 0.9 to 0.4 | Number of populations 100, crossover percentage (pc) 0.7 | Number of population $N = 100$, mutation rate 0.05, single point crossover |
| $C_1 = C_2 = 2$ | Mutation percentage (pm) 0.3, extra range factor for crossover ($\gamma$) 0.4 and mutation rate (mu) is 0.1 | Acceleration co-efficient ($C_1$) 1.5 and global learning coefficient ($C_2$) 2, $r1$ and $r2$ (0, 1), degree of importance: $\alpha1$ and $\alpha2$ (0.4, 0.4) and $\alpha3$ is 0.4 |

**TABLE 11.2**

Parameters Search Ranges [48] of RTC France Solar Cell Module

| PV System | RTC Solar Cell | |
|---|---|---|
| Parameter | Lower Limit | Upper Limit |
| $I_{ph}$ (A) | 0 | 1 |
| $I_{sd}$ ($\mu$A) | 0 | 1 |
| $R_s$ ($\Omega$) | 0 | 0.5 |
| $R_{sh}$ ($\Omega$) | 0 | 100 |
| $\eta$ | 1 | 2 |

For all calculations, the number of maximum iterations and population are set to 5000 and 100, respectively [53, 59, 60]. For a SDM and DDM search space we are limited to $\{I_{ph}, I_{sd}, R_s, R_{sh}, \eta\}$ and $\{I_{ph}, I_{sd1}, I_{sd2}, R_s, R_{sh}, \eta_1, \eta_2\}$. Hence we need an effective optimization tool so that it can find the optimal value from the search space of double diode and single diode PV cell models. The RTC solar cell parameter search range for the optimization is shown in Table 11.2.

Because of the stochastic nature of metaheuristics, in a given number of statements, it may give diverse output. This strategy includes performing each statistic 20 times for each case, after which the quantifiable study is finished using the data. For the modeling of solar cells, we used Matlab 2013b version, and the specification of the computer system is an Intel(R) Core (TM) i3processor, 4 GB RAM, with Windows7 operating system [61, 62]. In the course of these quantitative investigations, we tested and considered the veracity of the suggested analysis based on a number of models, including the computational efficiency test, the fitness test, the convergence test, the reliability test, and the accuracy test, which are all explained subsequently. Finally, a synopsis of the exhibitions is provided.

### 11.5.1.1 Fitness Test

Esteem output or wellness estimation of an advancement calculation is the most significant foundation to demonstrate its proficiency [59, 60]. Here, we have thought about three significant standards of fitness: worst fitness, mean, and best fitness obtained after multiple

**TABLE 11.3**

Comparative Study Based on Different Levels of Fitness

| Case | Method | Maximum RMSE | Minimum RMSE | Mean of RMSE |
|------|--------|--------------|--------------|--------------|
| | GA | 0.01753624 | 0.001234366 | 0.0025362 |
| RTC single diode | PSO | 0.00244805 | 0.001022083 | 0.0020571 |
| | HGAPSO | 0.00121203 | 0.000925481 | 0.0019254 |
| | GA | 0.03214202 | 0.001693715 | 0.0032191 |
| RTC double diode | PSO | 0.03602997 | 0.001184587 | 0.00236821 |
| | HGAPSO | 0.00291256 | 0.001025632 | 0.00156812 |

program runs. From Table 11.3, it may be seen that the proposed HGAPSO can arrive at the best wellness esteem (at least RMSE) for the entirety of the instances of photovoltaic frameworks. HGAPSO has the least RMSE for both RTC SDM and DDM model solar cells.

### 11.5.1.2 Reliability Test

An optimization tool ought to consistently reach the global minima as closely as conceivable; for example, it ought to be fruitful and effective in every run [63]. Thus, we tested the steadfast excellence of the hybrid HGAPSO in this subsection and make a comparative study with the other basic optimization techniques on the premise of standard deviation. The standard deviation estimates the changeability and consistency of the example or populace. Table 11.4 shows the proximity investigation dependent on the standard deviation. Standard deviations of HGAPSO are best for both the SDM and DDM models.

### 11.5.1.3 Computational Efficiency Test

Computational efficiency is the execution time taken by every optimization technique for determining the optimum value of the solar cell model [35]. Table 11.5 shows an average computation time for all the algorithms utilized in this research. From Table 11.5, when the parameters of the RTC double diode and single diode systems were improved, it was found that HGAPSO needed less computing time.

**TABLE 11.4**

Comparative Study Based on Standard Deviation

| Case | Method | Standard Deviation of RMSE |
|------|--------|----------------------------|
| | GA | 0.0088454 |
| RTC single diode | PSO | 0.02896407 |
| | HGAPSO | 0.00321841 |
| | GA | 0.00659481 |
| RTC double diode | PSO | 0.02646232 |
| | HGAPSO | 0.00321432 |

**TABLE 11.5**

Comparative Study Based on Computational Time

| Case | Method | Average Computational Time (Sec) |
|---|---|---|
| | GA | 78.8936 |
| RTC single diode | PSO | 169.4620 |
| | HGAPSO | 74.412 |
| | GA | 84.28 |
| RTC double diode | PSO | 294.37 |
| | HGAPSO | 68.643 |

### 11.5.1.4 Convergence Test

The precise statistical linkage cannot emerge from the searching performance of any optimization technique. An optimization technique is said to have better convergence speed when it provides the least RMSE for several runs with the variation of iteration number [56, 64]. Here we performed the convergence test by taking the iterations 100 to 5000, and each of the cases we ran 20 times. All the algorithms, HGAPSO, PSO, and GA, were run for both the models to obtain the minimum fitness. Figures 11.4 and 11.5 represent the convergence test for all the optimization tools. From the graphs, it has been concluded that HGAPSO gives a better convergence (both for the double diode model and single diode model) than the other two basic metaheuristic optimization techniques.

### 11.5.1.5 Accuracy Test

Accuracy test has been conducted to predict the simulated diode current with respect to the experimental current under the same experimental conditions [35, 65]. Here we use two different error indexes: relative error (RE) and individual absolute error (IAE), defined in Eqs. (11.18) and (11.19).

$$\text{IAE} = \left| I_{\text{measured}} - I_{\text{calculated}} \right| \tag{11.18}$$

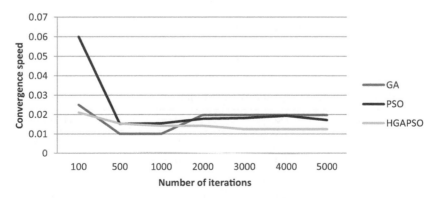

**FIGURE 11.4**

Convergence speed for single diode modeling using different metaheuristics.

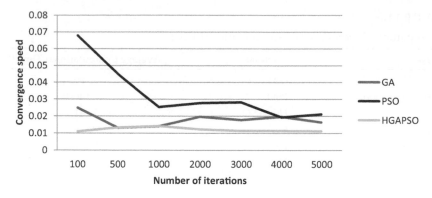

**FIGURE 11.5**
Utilizing various metaheuristics, convergence speed for modeling double diodes.

$$RE = \frac{I_{\text{measured}} - I_{\text{calculated}}}{I_{\text{measured}}} \tag{11.19}$$

Moreover, total absolute error ($TAE$) can be defined as

$$TAE = \sum_{i=1}^{n} IAE_i \tag{11.20}$$

Where $n$ is the number trial where $I_{\text{measured}}$ and $I_{\text{calculated}}$ are the calculated and experimental values of current. Tables 11.6 and 11.7 represent the optimum parameters of double diode and single diode PV cells by using three different optimization techniques. Tables 11.7 and 11.8 represent the total absolute error and accuracy of single diode and double diode models by utilizing the optimization techniques. From the experimental results it is seen that HGAPSO obtained the least RMSE and TAE.

### 11.5.2 Overall Efficiency

Presently we summarize the presentation of HGAPSO metaheuristics advancement system dependent on the previously mentioned assessment foundations and contrast other procedures. Consequently, we allocated an exhibition score for every calculation for every one of

**TABLE 11.6**

Optimal Parameters for Single Diode Model

| Parameters | PSO | GA | HGAPSO |
|---|---|---|---|
| $I_{\text{ph}}$ | 0.92 | 0.75 | 0.7233 |
| $I_{\text{sd}}$ | 0.452 | 0.4585 | 0.7109 |
| $R_s$ | 0.045 | 0.03985 | 0.00018 |
| $R_{\text{sh}}$ | 20.14 | 60.2308 | 10.33751 |
| $\eta$ | 1.8481 | 1.787 | 1.6222 |

**TABLE 11.7**

Optimal Parameters for Double Diode Model

| Parameters | PSO | GA | HGAPSO |
|---|---|---|---|
| $I_{\mathrm{ph}}$ | 0.73 | 0.7004 | 0.75978 |
| $I_{\mathrm{sd1}}$ | 0.928 | 0.59 | 053565 |
| $R_{\mathrm{s}}$ | 0.0039 | 0.0297 | 0.03449 |
| $R_{\mathrm{sh}}$ | 85.32 | 70.938 | 100 |
| $\eta_1$ | 1.7 | 1.582 | 1.53443 |
| $I_{\mathrm{sd2}}$ | 0.000126 | 0.235 | 0.125 |
| $\eta_2$ | 1.8 | 1.67 | 2 |

**TABLE 11.8**

Comparative Study Based on Total Absolute Error

| Case | Method | Total Absolute Error |
|---|---|---|
| RTC Single Diode | GA | 0.027807 |
| | PSO | 0.021422 |
| | HGAPSO | 0.01584 |
| RTC Double Diode | GA | 0.022134 |
| | PSO | 0.02431 |
| | HGAPSO | 0.01648 |

**TABLE 11.9**

Comparative Study Based on RMSE and Accuracy

| Case | Method | RMSE | Accuracy |
|---|---|---|---|
| RTC Single Diode | GA | 1.7861 | 98.214 |
| | PSO | 2.896 | 97.103 |
| | HGAPSO | 0.885 | 99.115 |
| RTC Double Diode | GA | 1.7689 | 98.2311 |
| | PSO | 2.646 | 97.354 |
| | HGAPSO | 0.926 | 99.074 |

the paradigms. The assessment of this score is based on the ratio of situations (capacities) when a computation achieves the best result (rule) to all other examples, such as single and dual diode demonstrations of RTC solar cells. Table 11.9 displays the close evaluation based on these results to evaluate the suggested HGAPSO calculation's productivity levels. The accompanying table illustrates how the recommended HGAPSO fared against various metaheuristics under all scenarios (Figures 11.6 and 11.7, Table 11.10).

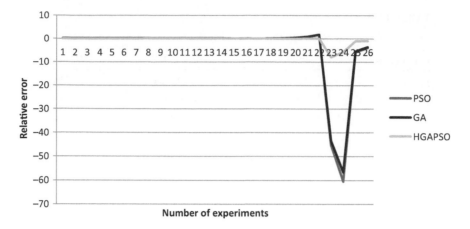

**FIGURE 11.6**
Relative error versus number of experimental data in single diode model using different metaheuristics.

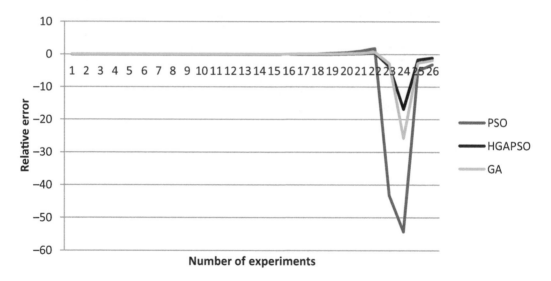

**FIGURE 11.7**
Relative error versus number of experimental data in double diode model using different metaheuristics.

### 11.5.3 Validation Between Manufacturer's Datasheet and Experimental Datasheets

In this section, to validate the proposed model, we extricate the optimum parameters for both two diode and single diode models of solar-based modules. The exploratory information is straightforward and helps to produce I-V characteristics which are identical to those given in the manufacturer datasheet at irradiance levels of 1000 W/m² and temperature of 25°C.

#### 11.5.3.1 Case Study 1: Single Diode Model

The optimal parameters of the single diode model are achieved when it gives the least RMSE after the 20 runs. Data sets have been taken for the temperature of 25°C and

**TABLE 11.10**

Comparative Study Based on Different Statistical Criteria

| Method | Standard Deviation | Time | Convergence | TAE | RE | RMSE | Accuracy | Max. RMSE | Min. RMSE | Mean RMSE | Total Score |
|---|---|---|---|---|---|---|---|---|---|---|---|
| GA | 0.33 | 0.33 | 0.5 | 0 | 0.5 | 0.33 | 0.33 | 0.33 | 0.33 | 0.33 | 3.31 |
| PSO | 0 | 0 | 0 | 0.33 | 0 | 0 | 0.33 | 0.33 | 0.33 | 0.33 | 1.65 |
| HGAPSO | 0.66 | 0.66 | 0.5 | 0.66 | 0.5 | 0.66 | 0.33 | 0.33 | 0.33 | 0.33 | 4.96 |

**TABLE 11.11**

The Extracted Optimum Parameters for S75 PV Module at 25°C Temperature and Irradiance Level of 1000 W/m² by HGAPSO (Single Diode Model)

| Solar Module | Temperature | $I_{ph}$ | $I_{sd}$ | $R_s$ | $R_{sh}$ | $\eta$ |
|---|---|---|---|---|---|---|
| S75 | 25°C | 4.27431 | 0.0771 | 0.2547 | 1939.26 | 1.481 |

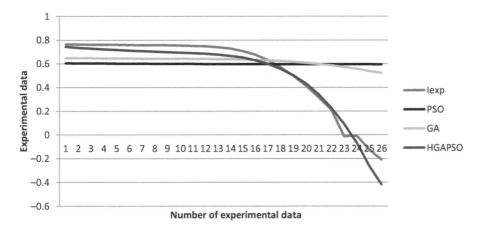

**FIGURE 11.8**
Comparative studies between experimental and calculated current for single diode model.

irradiance levels of 1000 W/m². Ideality factor $\eta$ for the ideal (indicated in literature) and the estimated single diode model in both cases lies in the interim [1, 2]. From Table 11.11, possible ranges of input parameters like series resistance fall inside the interim [0Ω, 1Ω]. The correlations between the assessed models and the test information of the single diode model at irradiance levels of 1000 W/m² are given in Figure 11.8.

### 11.5.3.2 Case Study 2: Double Diode Model

The optimal parameters of the two diode models are achieved when it gives the least RMSE after the 20 runs. Datasets have been taken for the irradiance levels of 1000 W/m² and temperature of 25°C. From Table 11.12, we obtained the possible ranges of input parameters, namely, that series resistance lies in [0 Ω, 0.06 Ω], parallel resistance is low (> 100 Ω), and

**TABLE 11.12**

The Extracted Optimum Parameters for S75 PV Module at 25°C Temperature and Irradiance Level of 1000 W/m² by HGAPSO (Double Diode Model)

| Solar Module | Temperature | $I_{ph}$ | $I_{sd1}$ | $I_{sd2}$ | $R_s$ | $R_{sh}$ | $\eta_1$ | $\eta_2$ |
|---|---|---|---|---|---|---|---|---|
| S75 | 25°C | 4.478 | 0.0000 | 0.0537 | 0.2843 | 7542.85 | 3.4522 | 1.1747 |

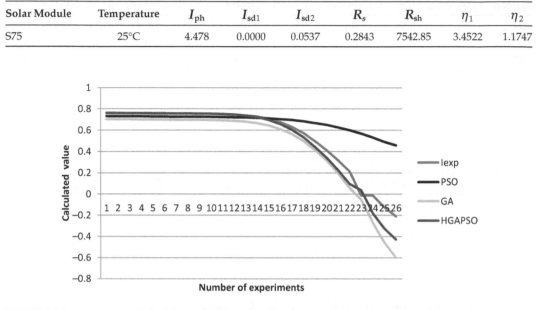

**FIGURE 11.9**
Comparative studies between experimental and calculated current double diode model.

ideal values of $\eta_1$ and $\eta_2$ would be 1 and 2 instead of industrial sample values of 1 and 5, respectively. The examinations between the evaluated models and the test information of the two diode models at irradiance levels of 1000 W/m² are given in Figure 11.9.

## 11.6 Conclusions

This paper presents optimal characteristics of the exploratory solar cell datasets of single diode and double diode RTC cells for the control and design of a PV system. The estimated current, as well as power rating, of a solar cell depends upon several input parameters; those that are mentioned are produced current, shunt resistance, series resistance, reverse saturation current, and ideality factor in Section 11.2. Due to the complexity of the optimization problem, a new multiobjective hybrid optimization method involving both PSO and GA is applied. Moreover, a comparison between the HGAPSO on the one hand, and PSO and GA each used in isolation on the other, is performed for both the single and double diode objective functions.

Due to its capability of searching for a global optimum and its convergence speed, the proposed HGAPSO algorithm is a fruitful algorithm that can be implemented as a supplementary method to assess the PV module simulation parameters. Section 11.5 detailed the proposed hybrid HGAPSO's efficiency against that of PSO and GA using convergence speed, computational efficiency, root mean square error, and accuracy.

A subsection of Section 11.5 explains the key study for verification of the effectiveness of HGAPSO with the experimental datasets and graphically. In this key study, the proposed hybrid algorithm is applied, precisely and proficiently retrieving the attributes of PV modules (S75) which have irradiance levels 1000 W/m$^2$ and temperature of 25°C.

Every segment of the results reveals that the suggested HGAPSO exhibitions surpassed the standard PSO and GA. However, a relatively large inaccuracy and standard deviation are among the major flaws of HGAPSO. Research should focus on enhancing the anticipated optimization's dependability and consistency. Presenting unique statement frameworks and modifying or aggregating the effectiveness might be a strategy to overcome these constraints.

## Conflict of Interest

According to the authors, there is no conflict.

## References

1. Sampaio PGV, González MOA. Photovoltaic solar energy: Conceptual framework. *Renewable and Sustainable Energy Reviews* 2017;74: 590–601.
2. Jia T, Dai Y, Wang R. Refining energy sources in winemaking industry by using solar energy as alternatives for fossil fuels: A review and perspective. *Renewable and Sustainable Energy Reviews* 2018;88: 278–296.
3. Ge TS, Wang RZ, Xu ZY, Pan QW, Du S, Chen XM, et al. Solar heating and cooling: Present and future development. *Renewable Energy* 2018;126: 1126–1140.
4. Bana S, Saini RP. A mathematical modeling framework to evaluate the performance of single diode and double diode based SPV systems. *Energy Reports* 2016;2: 171–187.
5. Rhouma MB, Gastli A, Brahim LB, Touati F, Benammar M. A simple method for extracting the parameters of the PV cell single-diode model. *Renewable Energy* 2017;113: 885–894.
6. Chan DS, Phang JC. Analytical methods for the extraction of solar-cell single-and double-diode model parameters from IV characteristics. *IEEE Transactions on Electron Devices* 1987;34: 286–293.
7. Ahmad T, Sobhan S, Nayan MF. Comparative analysis between single diode and double diode model of PV cell: Concentrate different parameters effect on its efficiency. *Journal of Power and Energy Engineering* 2016;4: 31–46.
8. Humada AM, Hojabri M, Mekhilef S, Hamada HM. Solar cell parameters extraction based on single and double-diode models: A review. *Renewable and Sustainable Energy Reviews* 2016;56: 494–509.
9. Derick M, Rani C, Rajesh M, Farrag ME, Wang Y, Busawon K. An improved optimization technique for estimation of solar photovoltaic parameters. *Solar Energy* 2017;157: 116–124.
10. Fathy A, Rezk H. Parameter estimation of photovoltaic system using imperialist competitive algorithm. *Renewable Energy* 2017;111: 307–320.
11. Kumari PA, Geethanjali P. Adaptive genetic algorithm based multi-objective optimization for photovoltaic cell design parameter extraction. *Energy Procedia* 2017;117: 432–441.
12. Olowu TO, Jafari M, Sarwat AI. A multi-objective optimization technique for volt-var control with high PV penetration using genetic algorithm. *Proceedings of the 2018 North American Power Symposium (NAPS)*. IEEE;2018, pp. 1–6.

13. Li S, Gu Q, Gong W, Ning B. An enhanced adaptive differential evolution algorithm for parameter extraction of photovoltaic models. *Energy Conversion and Management* 2020;205: 112443.
14. Li S, Gong W, Yan X, Hu C, Bai D, Wang L. Parameter estimation of photovoltaic models with memetic adaptive differential evolution. *Solar Energy* 2019;190: 465–474.
15. Huynh DC, Ho LD, Dunnigan MW. Parameter estimation of solar photovoltaic cells using an improved artificial bee colony algorithm. *Proceedings of the International Conference on Green Technology and Sustainable Development*. Springer;2020, pp. 281–292.
16. Premkumar M, Babu TS, Umashankar S, Sowmya R. A new metaphor-less algorithms for the photovoltaic cell parameter estimation. *Optik* 2020;208: 164559.
17. Premkumar M, Jangir P, Sowmya R, Elavarasan RM, Kumar BS. Enhanced chaotic JAYA algorithm for parameter estimation of photovoltaic cell/modules. *ISA Transactions* 2021;116: 139–166.
18. Xiong G, Zhang J, Shi D, Zhu L, Yuan X. Parameter extraction of solar photovoltaic models with an either-or teaching learning based algorithm. *Energy Conversion and Management* 2020;224: 113395.
19. Ramadan A, Kamel S, Korashy A, Yu J. Photovoltaic cells parameter estimation using an enhanced teaching–Learning-based optimization algorithm. *Iranian Journal of Science and Technology, Transactions of Electrical Engineering* 2020;44: 767–779.
20. Wang M, Zhang Q, Chen H, Heidari AA, Mafarja M, Turabieh H. Evaluation of constraint in photovoltaic cells using ensemble multi-strategy shuffled frog leading algorithms. *Energy Conversion and Management* 2021;244: 114484.
21. Sheng H, Li C, Wang H, Yan Z, Xiong Y, Cao Z, et al. Parameters extraction of photovoltaic models using an improved moth-flame optimization. *Energies* 2019;12: 3527.
22. Kanimozhi G, Kumar H. Modeling of solar cell under different conditions by ant lion optimizer with Lambert W function. *Applied Soft Computing* 2018;71: 141–151.
23. Aydin O, Gozde H, Dursun M, Taplamacioglu MC. Comparative parameter estimation of single diode PV-cell model by using sine-cosine algorithm and whale optimization algorithm. *Proceedings of the 2019 6th International Conference on Electrical and Electronics Engineering (ICEEE)*. IEEE;2019, pp. 65–68.
24. Mohamed M, Abdullah M, Yahia M. Parameters estimation of photovoltaic module using grey wolf optimization method. *Przegląd Elektrotechniczny* 2021;97.
25. Alghamdi MA, Khan MFN, Khan AK, Khan I, Ahmed A, Kiani AT, et al. PV model parameter estimation using modified FPA with dynamic switch probability and step size function. *IEEE Access* 2021;9: 42027–42044.
26. Ryad AK, Atallah AM, Zekry A. Photovoltaic parameters estimation using hybrid flower pollination with clonal selection algorithm. *Turkish Journal of Electromechanics Energy* 2018;3: 15–21.
27. Dutta P, Biswas SK, Biswas S, Majumder M. Parametric optimization of solar parabolic collector using metaheuristic optimization. *Computational Intelligence and Machine Learning* 2021;2: 26–32.
28. Mandal S. Modeling of photovoltaic systems using modified elephant swarm water search algorithm. *International Journal of Modelling and Simulation* 2020;40: 436–455.
29. Gupta J, Nijhawan P, Ganguli S. Parameter estimation of different solar cells using a swarm intelligence technique. *Soft Computing* 2021;26: 1–44. https://doi.org/10.21203/rs.3.rs-371558/v1
30. Dutta P, Agarwala R, Majumder M, Kumar A. Parameters extraction of a single diode solar cell model using bat algorithm, firefly algorithm & cuckoo search optimization. *Annals of the Faculty of Engineering Hunedoara* 2020;18: 147–156.
31. Mathew D, Rani C, Kumar MR, Wang Y, Binns R, Busawon K. Wind-driven optimization technique for estimation of solar photovoltaic parameters. *IEEE Journal of Photovoltaics* 2017;8: 248–256.
32. Dutta P, Kumar A. Design an intelligent flow measurement technique by optimized fuzzy logic controller. *Journal Europen Des Systems Automatics* 2018;51: 89–107.
33. Dutta P, Kumar A. Intelligent calibration technique using optimized fuzzy logic controller for ultrasonic flow sensor. *Mathematical Modelling of Engineering Problems* 2017;4: 91–94.

34. Dutta P, Kumar A. Study of optimized NN model for liquid flow sensor based on different parameters. *Proceedings of the International conference on Materials, Applied Physics & Engineering*. 2018.

35. Dutta P, Kumar A. Modelling of liquid flow control system using optimized genetic algorithm. *Statistics, Optimization & Information Computing* 2020;8: 565–582.

36. Sharma K, Cerezo M, Holmes Z, Cincio L, Sornborger A, Coles PJ. Reformulation of the no-free-lunch theorem for entangled datasets. *Physical Review Letters* 2022;128: 070501.

37. Li S, Gong W, Wang L, Yan X, Hu C. A hybrid adaptive teaching–learning-based optimization and differential evolution for parameter identification of photovoltaic models. *Energy Conversion and Management* 2020;225: 113474.

38. Li S, Gong W, Yan X, Hu C, Bai D, Wang L, et al. Parameter extraction of photovoltaic models using an improved teaching-learning-based optimization. *Energy Conversion and Management* 2019;186: 293–305.

39. Xiong G, Zhang J, Shi D, He Y. Parameter extraction of solar photovoltaic models using an improved whale optimization algorithm. *Energy Conversion and Management* 2018;174: 388–405.

40. Chen X, Yu K. Hybridizing cuckoo search algorithm with biogeography-based optimization for estimating photovoltaic model parameters. *Solar Energy* 2019;180: 192–206.

41. Liang J, Qiao K, Yu K, Ge S, Qu B, Xu R, et al. Parameters estimation of solar photovoltaic models via a self-adaptive ensemble-based differential evolution. *Solar Energy* 2020;207: 336–346.

42. Xiong G, Zhang J, Yuan X, Shi D, He Y, Yao G. Parameter extraction of solar photovoltaic models by means of a hybrid differential evolution with whale optimization algorithm. *Solar Energy* 2018;176: 742–761.

43. Oliva D, Abd El Aziz M, Hassanien AE. Parameter estimation of photovoltaic cells using an improved chaotic whale optimization algorithm. *Applied Energy* 2017;200: 141–154.

44. Ram JP, Babu TS, Dragicevic T, Rajasekar N. A new hybrid bee pollinator flower pollination algorithm for solar PV parameter estimation. *Energy Conversion and Management* 2017;135: 463–476.

45. Mughal MA, Ma Q, Xiao C. Photovoltaic cell parameter estimation using hybrid particle swarm optimization and simulated annealing. *Energies* 2017;10: 1213.

46. Beigi AM, Maroosi A. Parameter identification for solar cells and module using a hybrid firefly and pattern search algorithms. *Solar Energy* 2018;171: 435–446.

47. Dutta P, Majumder M, Kumar A. An improved grey wolf optimization algorithm for liquid flow control system. *International Journal of Engineering and Manufacturing* 2021;11: 10–21. https://doi.org/10.5815/ijem.2021.04.02.

48. Dutta P, Majumder M. An improved grey wolf optimization technique for estimation of solar photovoltaic parameters. *International Journal of Power and Energy Systems* 2021;41: 217–221.

49. Li S-F, Cheng C-Y. Particle swarm optimization with fitness adjustment parameters. *Computers & Industrial Engineering* 2017;113: 831–841.

50. Ghasemi M, Akbari E, Rahimnejad A, Razavi SE, Ghavidel S, Li L. Phasor particle swarm optimization: A simple and efficient variant of PSO. *Soft Computing* 2019;23: 9701–9718.

51. Mohanty S, Mishra A, Nanda BK, Routara BC. Multi-objective parametric optimization of nano powder mixed electrical discharge machining of AlSiCp using response surface methodology and particle swarm optimization. *Alexandria Engineering Journal* 2018;57: 609–619.

52. Bellubbi S, Mallick B, Hameed AS, Dutta P, Sarkar MK, Nanjundaswamy S. Neural network (NN)-based RSM-PSO multiresponse parametric optimization of the electro chemical discharge micromachining process during microchannel cutting on silica glass. *Journal of Advanced Manufacturing Systems* 2022;1–29. https://doi.org/10.1142/S0219686722500330.

53. Dutta P, Cengiz K, Kumar A. Study of bio-inspired neural networks for the prediction of liquid flow in a process control system. In *Cognitive big data intelligence with a metaheuristic approach*. Elsevier;2022, pp. 173–191.

54. Dutta P, Kumar A. Design an intelligent calibration technique using optimized GA-ANN for liquid flow control system. *Journal Européen Des Systèmes Automatisés* 2017;50: 449.

55. Dutta P, Majumder M, Kumar A. Parametric optimization of liquid flow process by ANOVA optimized DE, PSO & GA algorithms—Google search. *International Journal of Engineering and Manufacturing (IJEM)* 2021;11: 14–24.
56. Dutta P, Agarwala R, Majumder M, Kumar A. Parameters extraction of a single diode solar cell model using bat algorithm, firefly algorithm & cuckoo search optimization. *Annals of the Faculty of Engineering Hunedoara* 2020;18: 147–156.
57. Guvenc U, Duman S, Saracoglu B, Ozturk A. A hybrid GA-PSO approach based on similarity for various types of economic dispatch problems. *Elektronika Ir Elektrotechnika* 2011;108: 109–114.
58. Garg H. A hybrid PSO-GA algorithm for constrained optimization problems. *Applied Mathematics and Computation* 2016;274: 292–305.
59. Dutta P, Mandal S, Kumar A. Application of FPA and ANOVA in the optimization of liquid flow control process. *RCES* 2019;5: 7–11. https://doi.org/10.18280/rces.050102.
60. Dutta P, Mandal S, Kumar A. Comparative study: FPA based response surface methodology & ANOVA for the parameter optimization in process control. *Advances in Modelling and Analysis C* 2018;73: 23–27. https://doi.org/10.18280/ama_c.730104.
61. Dutta P, Paul S, Kumar A. Comparative analysis of various supervised machine learning techniques for diagnosis of COVID-19. In *Electronic Devices, Circuits, and Systems for Biomedical Applications*. Elsevier;2021, pp. 521–540.
62. Dutta P, Paul S, Obaid AJ, Pal S, Mukhopadhyay K. Feature selection based artificial intelligence techniques for the prediction of COVID like diseases. *Journal of Physics: Conference Series* 2021;1963: 012167.
63. Dutta P, Majumder M. Parametric optimization of drilling parameters in aluminum 6061T6 plate to minimize the burr. *International Journal of Engineering and Manufacturing* 2021;11(6): 36–47.
64. Dutta P, Kumar A. Design an intelligent calibration technique using optimized GA-ANN for liquid flow control system. *Journal Européen Des Systèmes Automatisés* 2017;50: 449–470. https://doi.org/10.3166/jesa.50.449-470
65. Mandal S, Dutta P, Kumar A. Modeling of liquid flow control process using improved versions of elephant swarm water search algorithm. *SN Applied Sciences* 2019;1: 1–16.

# 12

## Bio Inspired Optimization Based PID Controller Tuning for a Non-Linear Cylindrical Tank System

### 12.1 Introduction

In process industries, most of the real-time processing units, like biochemical reactors, cylindrical tank systems, and continuous stirred tank reactors (CSTRs), are highly nonlinear in nature; that is why tuning the controller parameter for these systems for stability as well as eliminating the disturbance is critical. So proper tuning of the controller parameter is essential in all the nonlinear process control systems.

Designing a controller parameter of a stable system is quite easy, but in a nonlinear process or unstable system, the designer should indicate the range (maximum and minimum) of controller parameters and the average of these limiting values as well to make the system stabilize. Increasing the time delay of the unstable system narrows the limiting value of the controller parameter to bring the system under control.

Researchers have used conventional controllers for the model based system [1], and according to their requirement, system models were further reduced to first order or first order with time delay systems. But for the unstable system, the fitting rule could not provide better results after reducing the system model. Most of the classical proportional + integral +differential (PID) tuning methods utilized computational techniques to get the best possible controller parameters. In recent years a number of heuristic algorithms have used a computational method for proper tuning of control parameters.

Most of the process industry has a vital task to maintain the liquid level in an organized manner. In the tanks, fluid is handled chemically or by transfusion, but liquid level in the tanks must be under stable conditions [2, 3]. In this process, the tank is chosen as a cylindrical shape in which liquid level is to be controlled. Cylindrical tank systems have great application in process industries such as chemical and pharmaceutical industries, food processing industries, and so on. Change in shape with respect to height makes a conical tank a nonlinear system. So, control of fluid is a major task [4], and an adaptive and reliable technique must be used to make the system stable.

There are a huge number of works done on PID tuning of a nondirect framework. S. M. Girirajkumar and D. Mercy et al. [5] have done their work tuning a PID controller utilizing a Z-N strategy and astute systems like the genetic algorithm. Lin et al. [6] have proposed receptive calculation for PID controllers dependent on a hypothesis of versatile collaboration. In Gole'a et al. [7], for nondirect frameworks where a multifarious law is gained by PI law, a fuzzy model reference flexible control has been proposed. In PID gain booking, Viljamaa and Koivo [8] have proposed a predictive argumentation framework. Marcelo et al. [9] have proposed a Lyapunov-based settling control structure technique for an

DOI: 10.1201/9781003216001-14

unsure nonstraight powerful framework utilizing as a fuzzy model. Chatterjee et al. [10] have proposed a near report between Z-N and fuzzy rationale based PID controllers for the speed restrictions of a DC engine. Pijush et al. [11, 12] proposed fuzzy logic based AI techniques for prediction of flow rate on contact types of flow sensor. Pijush et al. studied the performance of conventional AI techniques like ANN [13], ANFIS [14], and GA [15, 16] for process controller liquid level rate projection.

In this chapter, tuning the controller parameter of a cylindrical tank is done by the flower pollination algorithm (FPA) and bacteria foraging optimization (BFO). A detailed study of the different measurement of error and transient properties of both the transfer functions (first-order system and first-order system with time delay) of a cylindrical tank are analyzed by both the algorithms.

The rest of the chapter is organized as follows. Section 12.2 exhibits the overall methodology of the research, which elaborates a diagram displaying a tube shaped tank and a brief portrayal of FPA and BFO based controllers. Section 12.3 shows the simulated outcomes on BFO and FPA process models and a continuous execution utilizing a cylindrical tank framework. Segment 4 concludes the present research work.

## 12.2 Methodology

Intelligent control mechanism iterative methods determine the proportional, integral, and differential transformation functions for using the error and derivation erroneous inputs, and then uses these results to update the controller gains of PID controllers. Three distinct tweaking criteria—integrated absolute error (IAE),integral of square error (ISE), and integral time absolute error (ITAE)—were used to evaluate which of the controllers' skill level was optimal. These defect integrals each reflect the extent and nature of an imperfection as a type of correction factor. Again the performance of PID controller depends on proportional gain ($k_p$), integral gain ($k_i$), and derivative gain ($k_d$). These gains can be got by using the following methods: Ziegler-Nichols (ZN), gain-phase margin, root locus, minimum variance, and gain scheduling. However, some of these strategies are rather difficult, and they are not the best for manipulating unpredictable, high-order systems. Different search techniques are given to enhance the efficacy and get appropriate $k_p, k_i$, and $k_d$. Numerous methods have been suggested in some research for assessing PID computational efficiency: ITAE, integral of timing-weighted-squared-error, and integrated absolute error (IAE) performance evaluation (ITSE).

In this research, the overall process was performed in three stages as shown in Figure 12.1. In stage 1, the mathematical formulation of a cylindrical tank system is formulated by the inflow rate and outflow rate, which are described in Subsection 12.2.1. In stage 2, the objective function is obtained with the help of PID controller transfer function along with the transfer function of the cylindrical tank. Finally two different bio-inspired optimization techniques are applied for finding the optimum value of proportional gain ($k_p$), integral gain ($k_i$), derivative gain ($k_d$), and error indices.

### 12.2.1 Mathematical Model of Cylindrical Tank

For the field of pharmaceutical industries, a cylindrical tank is used due to laminar flow of liquid according to the flow regime (Reynolds number is less than 2000). The mathematical

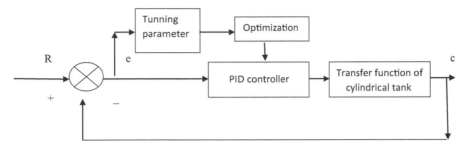

**FIGURE 12.1**
Structure of the proposed model.

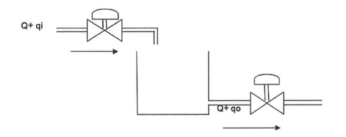

**FIGURE 12.2**
Schematic diagram of the conical tank system [17].

model of a cylindrical tank is designed by a differential equation utilized on the concept of resistance and capacitance. Figure 12.2 shows the schematic diagram of the cylindrical tank [17].

Operating parameters of cylindrical tank

$q_i$ = small deviation of inflow rate from its steady-state value

$q_0$ = small deviation of outflow rate from its steady-state value

$H$ = steady-state head

$h$ = small deviation of head from its steady-state value

If we assume that $q_i$ is the platform's input and $h$ is its output, the transmission design of the system is obtained by

$$\frac{H(s)}{Qi(s)} = \frac{R}{(1+RCS)} \tag{12.1}$$

If $q_0$ is taken as the output and the input is $q_i$, then the transfer function of the system is given as follows:

$$q_0 = \frac{H}{R}$$

Taking the Laplace transform of both the sides, we get

$$Q0(s) = \frac{H(s)}{R} \qquad (12.2)$$

From Eqs (12.11) and (12.12), we obtain [17]

$$\frac{Q0(s)}{Qi(s)} = \frac{R}{(1+RCS)} \qquad (12.3)$$

Where $R = 0.2$, $C = 30.4$

After reckoning the $t1$ and $t2$, the time delay $(= \theta)$ and the process time constant $(= T)$ can be obtained by the consecutive equation $\theta = 1.3t1 - 0.29\ t2 = 0.525$, $T = 0.67(t2-t1) = 6.1841$

Now the transfer function can be represented by the equivalent first order time delay form is [17]

$$G(s) = \frac{K}{(Ts+1)} \exp(-\theta * s)$$

$$G(s) = \frac{0.23}{(6.18s+1)} \exp(-0.525 * s) \qquad (12.4)$$

## 12.2.2 Description of Metaheuristic Techniques

### 12.2.2.1 Flower Pollination Algorithm (FPA)

The flower pollination algorithm is a cosmos-obeying algorithm discovered by Xin-She Yang in 2012. In FPA, the pollination process occurs through some carriers like wind, insects, birds, bats, and so on. Flower pollination basically deals with pollen transference from a flower's male organ to its female part [18] via wind, water, or insects or other animals. The pollination process basically deals with the young plants' propensity to propagate. The pollination process can be of two types: biotic and abiotic processes. Biotic pollination processes include livings pollinators such as birds, insects, and so on to maneuver pollen from one flower to another. Abiotic pollination involves nonliving pollinators like wind, water, and so on to transfer pollen [19, 20]. It has been surveyed that 90% of the flowering plants deals with the biotic pollination, which needs pollinators for reproduction of plants, and about 10% occurs without any pollinator leading to abiotic forms of pollination. There are two methods in pollination, namely self-pollination and cross-pollination. Self-pollination arises from pollen of the identical flower or various bloomings of the same plant without the aid of any pollinator, whereas cross-pollination uses pollen from a flower of a different strain [21]. Flower constancy is a process in which pollinators only visit certain types of flower plant species and increase the identical flower species reproduction. FPA relies on the following four rules for pollination [21]:

1. Pollen-carrying carriers can accompany dispersion flights and can be seen as part of a global pollination process together including biological and cross-pollination.
2. Local pollination comprises both biotic and self-pollination.

3. Flower interdependence is defined as the expectation of reproduction being equivalent to the similarity of two entwined blooms.

4. A switch probability of p [0, 1] maintains both the local and global pollination situations. Local pollination may contribute significantly in the entire pollination process, in addition to physical intimacy and other factors like wind and water. In FPA, two fundamental plans, to be specific global fertilization and local fertilization processes, are there. Worldwide pollination includes the dusts that are moved by insects, birds, bats, and so on to an enormous separation because of they can fly longer range. In this procedure, there is likelihood to a get natural selection g*[best solution]. The worldwide fertilization can be composed numerically as

$$X_i^{t+1} = X_i^t + L(\lambda)(X_i - g^*) \tag{12.5}$$

Here, $X_i^t$ is the dust $i$ or arrangement vector $X_i$ at emphasis $t$, $g^*$ is the present best arrangement among all arrangements in the current cycle and $L(\lambda)$ is a stage size. Since transporters might be moved over a bigger zone with various separation steps; L'evy flight [22] is utilized to express this wonder. For $L > 0$, L'evy appropriation can be composed as

$$L \sim \frac{\lambda \Gamma(\lambda) \sin(\pi\lambda/2)}{\pi} \frac{1}{s^{\lambda+1}} \tag{12.6}$$

Here $\Gamma(\lambda)$ is the standard gamma capacity, and L'evy appropriation is valid for longer advances $S > 0$. Local fertilization is communicated scientifically utilizing Rule 2 and Rule 3 as

$$X_i^{t+1} = X_i^t + \varepsilon(X_j^t - X_k^t) \tag{12.7}$$

Here $X_j^t$ and $X_k^t$ are dust from discrete blossoms of the indistinguishable plant species. In the event that $X_j^t$ and $X_k^t$ results from the indistinguishable species or looked over the indistinguishable populace, this is practically like a neighborhood arbitrary walk if a diagram can be drawn from a uniform conveyance in [0, 1]. In modern research FPA is useful in different research domains [23–25]: process control optimization, renewable energy, the chemical industry, biomedical field, and so on (Table 12.1).

**TABLE 12.1**

Parameter of FPA

| | |
|---|---|
| Population size | 20 |
| Probability switch | 0.8 |
| Number of iteration | 20–100, interval of 20 |
| Number of the search variable | 3 |
| Dimension of the search | −25 to 25 |

### 12.2.2.2 Bacterial Foraging Optimization Algorithm (BFOA)

BFOA is also a nature-obeying algorithm, developed by Kevin M. Passino in 2002. It is a mechanism based on biological encouragement that mimics the foraging activities of *Escherichia coli* (*E. coli*) bacteria resides in the human intestine. The objective of this algorithm is to eliminate weak foraging bacteria and keep the strong foraging bacteria to maximize energy per time. The *E. coli* bacterium consists of a plasma membrane, a cell wall, and a capsule containing cytoplasm and nucleoid. The locomotion is done with the help of flagella. The bacteria can travel in two different ways, namely tumbling and swimming [26] using flagella. When flagella rotate in the clockwise direction, each flagellum pulls the cell and finally the bacterium tumbles. When flagella rotate in the anticlockwise direction, each flagellum pushes the cell to cause the bacterium to swim at a faster rate in search for food [27]. These two processes continue alternately to move the bacterium in search of nutrients at different directions. Bacteria can interact with each other by sending various signals. With the help of tumbling and swimming, the bacteria can travel a longer distance for higher concentrations of food and to avoid harmful places. There are four major steps in the BFO algorithm [28] as follows:

*Chemotaxis*: Chemotaxis is the fundamental strides of the microscopic organisms' scrounging procedure. The *E. coli* bacterium can go in two different ways, swimming and tumbling. In swimming, the bacterium swims a single way to gather nourishment, and in tumbling, the bacterium alters its course to another for nutrient gradient. Accepting $\theta^i(j,k,l)$ shows the *i*-th bacterium at *j*-th chemo strategy step, *k*-th proliferation step, and *l*-th disposal and dispersal step. The portability of the bacterium is spoken to as follows:

$$\theta^i(j+1,k,l) = \theta^i(j,k,l) + C(i)\frac{\Delta(i)}{\sqrt{\Delta^T(i)\Delta(i)}} \tag{12.8}$$

Here, $C(i)$ is the step size during each swim or tumble and $\Delta(i)$ represents a random vector whose elements lie in $[1,-1]$.

*Swarming*: A group of microscopic organisms helps organizes themselves into a revolving ring by ascending the supplementary tilt. If the stress response is triggered by a higher level of short, they release an attractant suction that induces them to congregate and travel as concentric samples of swarming with a couple of layers of bacteria. The swarming process is spoken to scientifically as follows:

$$J_{CC}\left(\theta, P(j,k,l)\right) = \sum_{i=1}^{s} j_{cc}(\theta,\theta^i(j,k,l)) = \sum_{i=1}^{s}\left[-d_{\text{attractant}}\exp(-w_{\text{attractant}}\right.$$

$$\left.\sum_{m=1}^{p}\left(\theta_m-\theta_m^i\right)^2\right] + \sum_{i=1}^{s}\left[h_{\text{repellant}}\exp\left(-w_{\text{repellant}}\sum_{m=1}^{p}\left(\theta_m-\theta_m^i\right)^2\right)\right] \tag{12.9}$$

Here, $J_{CC}(\theta,P(j,k,l))$ is the objective function value to be joined to the original objective function. $S$ is the total number of bacteria, $p$ is the number of variables to be optimized and $\theta = [\theta_1, \theta_2,...,\theta_p]^T$ is a point in the *p*-dimensional search domain. $d_{\text{attractant}}$, $w_{\text{attractant}}$, $h_{\text{repellant}}$, $w_{\text{repellant}}$ are different coefficients that should be used properly [28].

*Reproduction*: After these two steps some bacteria have a good amount of food and some bacteria have less food. Those bacteria which have enough food will survive, and the

**TABLE 12.2**

Parameters of BFO

| | |
|---|---|
| S: Number of bacteria | 19 |
| D: Number of parameters to be optimized | 3 |
| NS: Swimming length after which tumbling of bacteria | 3 |
| Nre: Maximum number of reproductions to be undertaken | 6 |
| Ned: Maximum number of elimination–dispersal events | 2 |
| Ped: Probability with which the elimination–dispersal will continue | 0.25 |

rest of them will die. In this step, healthier bacteria divided into two bacteria which are then moved in the sameness location. This tracks the size of the swarm fixed, and healthier bacteria keep reproducing.

*Elimination and Dispersal*: When the temperature of the high nutrient gradient area suddenly increased, then all the bacteria in that area are destroyed, or a bunch is dispersed into a new area. This process relocates the bacteria to a new place to escape a noxious environment (Table 12.2).

## 12.3 Results and Discussion

$$ITAE = \int_0^\infty t|e(t)|dt \tag{12.10}$$

$$IAE = \int_0^\infty |e(t)|dt \tag{12.11}$$

$$MSE = \frac{1}{t}\int_0^\infty t[e(t)]^2 dt \tag{12.12}$$

**TABLE 12.3**

PID Tuning Parameters by BFO Algorithm for Ordinary Transfer Function

| Function | Number of Iterations | $k_p$ | $k_d$ | $k_i$ |
|---|---|---|---|---|
| $T(s) = \dfrac{0.23}{7s+1}$ | 20 | 7.7939 | 7.1118 | 4.3870 |
| | 40 | 6.4854 | 2.2700 | 2.3294 |
| | 50 | 8.4119 | 5.7796 | 2.0940 |
| | 60 | 4.9076 | 3.9160 | 1.4720 |
| | 80 | 8.6776 | 5.8946 | 1.9139 |
| | 100 | 6.8421 | 7.3495 | 1.8635 |

**TABLE 12.4**

Optimal Value and Elapsed Time of BFO for Ordinary Transfer Function

| Function | Number of Iterations | Optimal Best Fitness Value | Elapsed Time |
|---|---|---|---|
| $T(s) = \dfrac{0.23}{7s+1}$ | 20 | 49.7776 | 174.063899 |
| | 40 | 21.0245 | 351.513075 |
| | 50 | 22.2232 | 457.636032 |
| | 60 | 15.5019 | 382.295558 |
| | 80 | 22.2456 | 463.506272 |
| | 100 | 17.8902 | 594.390941 |

**TABLE 12.5**

PID Tuning Parameter by FPA Algorithm for Ordinary Transfer Function

| Function | No of Iteration | $k_p$ | $k_d$ | $k_i$ |
|---|---|---|---|---|
| $T(s) = \dfrac{0.23}{7s+1}$ | 20 | 17.9615 | −11.3237 | 1.1328 |
| | 40 | 16.5188 | −25.0000 | 1.0868 |
| | 50 | 9.3039 | −25.0000 | 1.0097 |
| | 60 | 24.6658 | −8.4462 | 1.1087 |
| | 80 | 17.2079 | −25.0000 | 1.0179 |
| | 100 | 8.9698 | −25.0000 | 1.2150 |

**TABLE 12.6**

PID Tuning Parameter by FPA Algorithm for Ordinary Transfer Function

| Function | No of Iteration | Optimal Best Fitness Value | Elapsed Time |
|---|---|---|---|
| $T(s) = \dfrac{0.23}{7s+1}$ | 20 | 1.311 | 29.954760 |
| | 40 | 0.56249 | 60.14014 |
| | 50 | 1.8654 | 74.006416 |
| | 60 | 1.2802 | 105.587651 |
| | 80 | 1.0123 | 127.323537 |
| | 100 | 2.6779 | 153.72954 |

**TABLE 12.7**

PID Tuning Parameters by FPA Algorithm for FOPD Transfer function

| Function | Number of Iteration | $k_p$ | $k_d$ | $k_i$ |
|---|---|---|---|---|
| $T(s) = \dfrac{0.23}{6.18s+1}e^{-0.525s}$ | 20 | −13.1299 | −1.0278 | 18.5636 |
| | 40 | 22.1051 | 11.6880 | −12.5917 |
| | 50 | −20.9579 | 16.1479 | 19.6661 |
| | 60 | −25.0000 | −11.1245 | −0.5083 |
| | 80 | 3.4650 | −7.7788 | −18.7215 |
| | 100 | 20.1734 | −0.0510 | −1.0162 |

**TABLE 12.8**

PID Tuning Parameters by FPA Algorithm for FOPD Transfer Function

| Function | Number of Iterations | Optimal Best Fitness Value | Elapsed Time |
|---|---|---|---|
| $T(s) = \dfrac{0.23}{6.18s+1}e^{-0.525s}$ | 20 | 123.22 | 33.757265 |
| | 40 | 98.4234 | 108.490897 |
| | 50 | 88.3684 | 162.032953 |
| | 60 | 96.9638 | 143.382602 |
| | 80 | 89.2301 | 119.038884 |
| | 100 | 50.8714 | 148.041570 |

**TABLE 12.9**

PID Tuning Parameter by BFO Algorithm for FOPD Transfer Function

| Function | No of Iteration | $k_p$ | $k_d$ | $k_i$ |
|---|---|---|---|---|
| $T(s) = \dfrac{0.23}{6.18s+1}e^{-0.525s}$ | 20 | 30.5040 | 10.8062 | 42.2983 |
| | 40 | 26.9336 | 27.3297 | 34.8356 |
| | 50 | 19.4059 | 17.4537 | 33.1396 |
| | 60 | 27.3052 | 13.2623 | 16.6599 |
| | 80 | 8.8490 | 21.0055 | 27.9986 |
| | 100 | 19.6958 | 27.7809 | 36.5638 |

**TABLE 12.10**

Optimal Value and Elapsed Time of BFO for first order plus dead time (FOPD) Transfer Function

| Function | Number of Iterations | Optimal Best Fitness Value | Elapsed Time |
|---|---|---|---|
| $T(s) = \dfrac{0.23}{6.18s+1}e^{-0.525s}$ | 20 | 21.2406 | 189.63421 |
| | 40 | 21.0245 | 244.562542 |
| | 50 | 20.9767 | 292.31456 |
| | 60 | 23.4052 | 338.935617 |
| | 80 | 21.0680 | 494.977125 |
| | 100 | 20.9725 | 622.434979 |

**TABLE 12.11**

Comparative Study for Time Domain Analysis

| Parameter | BFO-Ordinary (100) | FPA-Ordinary | BFO-FOPD | FPA-FOPD |
|---|---|---|---|---|
| Rise Time | 8.9030 | 23.1515 | NaN | 9.0639e-04 |
| Settling Time | 26.0901 | 42.4131 | NaN | 0.1409 |
| Settling Min | 3.8326 | 7.3878 | NaN | −2.2216 |
| Settling Max. | 4.4680 | 8.2053 | NaN | 0.4844 |
| Overshoot | 6.1719 | 0 | NaN | 128.0934 |
| Undershoot | 0 | 15.1753 | NaN | 506.7408 |
| Peak | 4.4680 | 8.2053 | Inf | 2.2216 |
| Peak Time | 18.0380 | 109.4433 | Inf | 0.0177 |

**TABLE 12.12**

Comparative Study for Error Indices after 100 Iteration

| Name of the system | ITAE | IAE | MSE |
|---|---|---|---|
| BFO-Ordinary | 93.249 | 5.17 | 26.728 |
| BFO-FODP | inf | inf | inf |
| FPA-Ordinary | 109.44 | 14.17 | 200.78 |
| FPA-FODP | 2.249 | 127.09 | 1.61*E04 |

## 12.4 Conclusion

Because of perplexing and dynamic procedures in the field of pharmaceuticals, they frequently have evil conditions because of an absence of capacity in the creation of the procedure control framework and lack of control execution. This investigation intends to upgrade the exhibition of proportional + integral + differential (PID) control of the yield stream rate from numerical display and hunt for ideal focuses. Optimization of PID control tuning parameters FPA and BFO to deal with nonlinear frameworks with evaporator undershoot reaction qualities that are hard to treat and improve the framework reaction with overshoot, where rise time is very long. Overall analysis is done for the different models, the ordinary transfer model and the first order model with time delay.

For PID tuning of the ordinary transfer function of a cylindrical tank, the flower pollination algorithm is superior to the bacteria foraging in terms of optimum value of the objective function (Figure 12.3) and elapsed time, but some of the transient response of the system, like rise time and settling time, is comparatively trailing the bacteria foraging optimization. For PID tuning of the first order time delay transfer function of a cylindrical tank, the flower pollination algorithm is superior in terms of elapsed time and transient parameters while bacteria foraging has less optimum value in terms of the objective activity (in Figure 12.4). Therefore,, FPA is better than BFO in terms of ITAE and MSE for PID tuning of the transfer function of cylindrical tank.

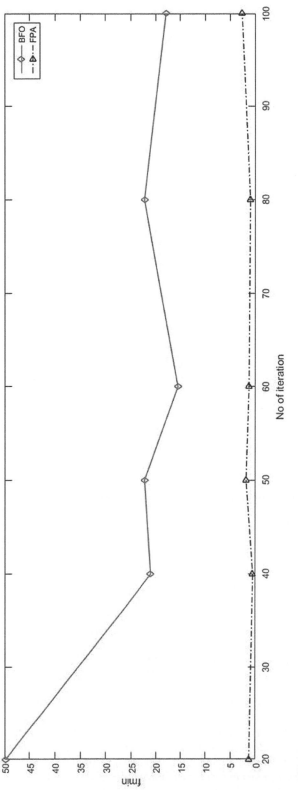

**FIGURE 12.3**
Optimal value of BFO and FPA for ordinary transfer function.

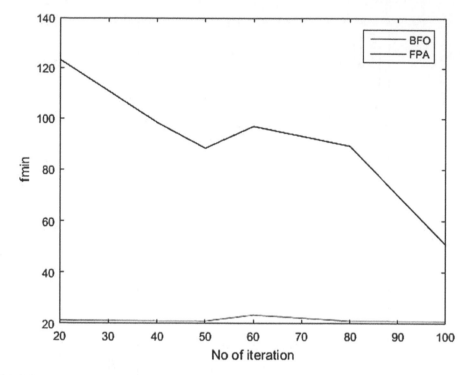

**FIGURE 12.4**
Optimal values of BFO and FPA for FOPD Transfer Function.

## References

1. Dutta P, Kumar A. Comparison of PID controller tuning techniques for liquid flow process control. *Global Journal on Advancement in Engineering and Science (GJAES)* 2016;2(1): 18–25.
2. Tripathi A, Verma RL, Alam MS. Design of Ziegler Nichols tuning controller for AVR system. *International Journal of Research in Electronics & Communication Technology* 2013;1(2): 154–158.
3. Astrom KJ, Hagglund T. *PID controllers: Theory, design, and tuning, instrument society of America* (2nd ed.). Research Triangle Park;1995, pp. 1–354.
4. Bhuvaneswari NS, Uma G, Rangaswamy TR. Adaptive and optimal control of a non-linear process using intelligent controllers. *Applied Soft Computing* 2009;9(1): 182–190.
5. Mercy D, Girirajkumar SM. Tuning of controllers for non linear process using intelligent techniques. *International Journal of Advanced Research in Electrical, Electronics and Instrumentation Engineering* 2013;2(9): 4410–4419.
6. Lin F, Brandt RD, Saikalis G. Self-tuning of PID controllers by adaptive interaction. *Proceedings of the 2000 American Control Conference. ACC (IEEE Cat. No. 00CH36334).* IEEE;2000; Vol. 5, pp. 3676–3681.
7. Golea N, Golea A, Benmahammed K. Fuzzy model reference adaptive control. *IEEE Transactions on Fuzzy Systems* 2002;10(4): 436–444.
8. Viljamaa P, Koivo HN. Fuzzy logic in PID gain scheduling. *Proceedings of the 3rd European Congress on Fuzzy and Intelligent Technologies EUFIT'95.* 1995, pp. 927–931.
9. Teixeira MCM, Zak SH. Stabilizing controller design for uncertain nonlinear systems using fuzzy models. *IEEE Transactions on Fuzzy Systems* 1999;7(2): 133–142.

10. Chatterjee S, Sil P, Dutta P. Comparative study Between the Z-N& Fuzzy PID controller for the speed control of a DC motor. *Proceedings of the CON 2016, 3rd National Conference on Emerging Trends in Science, Technology & Management for National Development*. 18–19 March, 2016.

11. Dutta P, Kumar A. Intelligent calibration technique using optimized fuzzy logic controller for ultrasonic flow sensor. *Mathematical Modelling of Engineering Problems* 2017;4(2): 91–94.

12. Dutta P, Kumar A. Design an intelligent flow measurement technique by optimized fuzzy logic controller. *Journal Européen des Systèmes Automatisés* 2018;51(1–3): 89.

13. Dutta P, Asok K. Study of optimized NN model for liquid flow sensor based on different parameters. *Proceeding of International Conference on Materials, Applied Physics and Engineering*, 2018.

14. Dutta P, Kumar A. Application of an ANFIS model to optimize the liquid flow rate of a process control system. *Chemical Engineering Transactions* 2018;71: 991–996.

15. Dutta P, Kumar A. Modelling of liquid flow control system using optimized genetic algorithm. *Statistics, Optimization & Information Computing* 2020;8(2): 565–582.

16. Dutta P, Kumar A. Design an intelligent calibration technique using optimized GA-ANN for liquid flow control system. *Journal Européen des Systèmes Automatisés* 2017;50(4–6): 449.

17. Dutta P, Kumar A. A study on performance on different open loop PID tuning technique for a liquid flow process. *International Journal of Information Technology, Control and Automation (IJITCA)* 2016;6: 2. https://doi.org/10.5121/ijitca.2016.6202

18. Walker M How flowers conquered the world. *BBC Earth News*. 2009, pp. 1–10.

19. Kaur G, Singh D. Pollination based optimization for color image segmentation. *International Journal of Computer Engineering & Technology* 2012;3(2): 407–414.

20. Kumar S, Bansal S, Bhalla P. Optimal Golomb ruler sequence generation for FWM crosstalk elimination: A BB–BC approach. *Proceedings of 6th International Multi Conference on Intelligent Systems, Sustainable, New and Renewable Energy Technology and Nanotechnology (IISN–2012), Institute of Science and Technology Klawad–133105, Haryana, India;*2012, pp. 255–262.

21. Yang XS. Flower pollination algorithm for global optimization. *International Conference on Unconventional Computing and Natural Computation*. Springer, Berlin, Heidelberg;2012, pp. 240–249.

22. Pavlyukevich I. Lévy flights, non-local search and simulated annealing. *Journal of Computational Physics* 2007;226(2): 1830–1844.

23. Dutta P, Kumar A. Modeling and optimization of a liquid flow process using an artificial neural network-based flower pollination algorithm. *Journal of Intelligent Systems* 2020;29(1): 787–798.

24. Dutta P, Mandal S, Kumar A. Comparative study: FPA based response surface methodlgy and ANOVA for the parameter optimization in process control. *Advances in Modelling and Analysis C* 2018;73(1): 23–27.

25. Dutta P, Mandal S, Kumar A. Application of FPA and ANOVA in the optimization of liquid flow control process. *Review of Computer Engineering* 2018;5(1): 7–11.

26. Berg HC, Brown DA. Chemotaxis in escherichia coli analysed by three-dimensional tracking. *Nature* 1972;239(5374): 500–504.

27. Abraham A, Hassanien AE, Siarry P, Engelbrecht A. *Foundations of computational intelligence Volume 3: Global optimization*. Springer;2009, Vol. 203.

28. Passino KM. Biomimicry of bacterial foraging for distributed optimization and control. *IEEE Control Systems Magazine* 2002;22(3): 52–67.

# 13

## A Hybrid Algorithm Based on CSO and PSO for Parametric Optimization of Liquid Flow Model

### 13.1 Introduction

Over the past decades, optimization has played a key role in the process industry, with artificial intelligence techniques assuming a leading role in development of the process model [1]. The modern process industry uses AI to control the liquid flow rate, liquid level, mixing of chemical products, and other many possible applications. Generally, the optimization technique used to find the optimal response corresponds to a given set of input variables on any nonlinear model [2]. The liquid flow control process is one of most common nonlinear models, where the response variable of flow rate relies upon a number of influencing attributes like type of sensor, characteristics of liquid, surrounding temperature, and so on. Due to complex relations between multivariable inputs and response, and a large delay time, this process model has limitations for conventional optimization techniques [3].

In most of the nonlinear complex process models, researchers have adopted computational optimization as it takes less computational time and predicts the potential input attributes correctly [4, 5]. Performance of computational optimization is generally segmented into two parts. During the first phase, a model was designed with the help of a majority number of datasets containing input and output variables. In the second phase, the model was validated and tested against a test dataset which accommodates the input and output process variables [6].

A number of researches have been carried out by many AI techniques for the parametric optimization and design of the gray box model on the basis of liquid flow control process; some these are the neural network control model [7, 8], fuzzy logic controller [9, 10], genetic algorithm [11–13], hybrid GA-ANN model [14, 15], ANFIS model [16, 17], and so on. Empirical models like analysis of variance (ANOVA), response surface methodology (RSM), and ANN are used to represent any nonlinear process plant training dataset, and finally metaheuristic optimization is used to find the optimal process parameters so that the test model is best fitted with the experimental dataset. There have been several researches performed for parametric optimization of empirical models by different metaheuristic optimization techniques for the liquid flow process industry; some of them are FPA-ANN [18], FPA-ANOVA [19], FPA-RSM [20], and improved versions of the original elephant swarm water search algorithm (ESWSA) based ANOVA and RSM model [21]. In addition to these, to improve convergence speed, a hybrid particle swarm optimization and grey wolf optimization (HPSOGWO) has also been proposed to implement this complex model [22]. The entire AI model operated within the constraint boundary conditions and fluctuation of complex features. The outcomes might yet be improved, though.

DOI: 10.1201/9781003216001-15

Therefore, we still have an issue with estimation of a super effective model for defining a flow rate control process. Particle swarm optimization is a very fruitful algorithm for finding the solution of any nonlinear objective function with the main aim to reduce the computational time and transfer time, and it also depicts a better solution and runs faster than other metaheuristic algorithms. However, its solution may become trapped in local optima, and due to fast convergence its results may not be accurate. On the other hand, the genetic algorithm provides the global optimum solution, but it requires a significant amount of computational time as it provides the better result by increasing the number of iterations as well as number of computational steps. Hence in this research we proposed a hybrid optimization algorithm, HGAPSO, by taking the advantage of both PSO and GA.

Comparatively to other strategies with the same objectives, the HGAPSO technique is projected to function faster with varied sizes of workflow applications [23–26]. Furthermore, because the GA mutation operator is used to improve the accuracy of the solutions found for many complicated and nonlinear problems, the HGAPSO algorithm may not get caught in the local optimal solution.

It is possible to draw the conclusion that the optimization tactics are effective based on the simulation's results that have been provided being both efficient and practical for achieving the real influence needs of the liquid flow monitoring procedure. The remainder of this essay is arranged as follows: After an orientation, Section 13.3 provides a quick emergence with the quantitative explanation of modeling of the liquid flow control mentioned in Section 13.2. The intended approach is explained in Section 13.4; the findings and discourse are then expressed; and finally, implications are provided in Sections 13.5 and 13.6, respectively.

## 13.2 Experimental Setup Liquid Flow Control Process

The flow and level measurement and control setup [10] (Figure 13.1, model number WFT-20-I) can be used to perform the research. In all, it provided 134 random samples from which four hypotheses—pipe diameter, output voltage, viscosity, and liquid (water) conductivity—have been factored into the equation for this investigation.

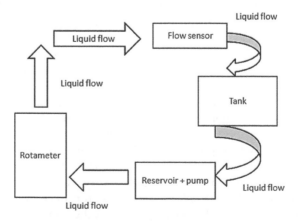

**FIGURE 13.1**
Experimental setup for liquid flow rate measurement [10].

## 13.3 Modeling of the Liquid Flow Process

It is particularly complicated to ascertain the procedure input factors, such as change in liquid properties and pipe diameter, to accomplish the best possible liquid flow rate and level using traditional controller approaches because of the nonlinear features of the liquid flow rate and liquid level processing. In order to optimize complicated mathematical models like ANOVA [11, 27] and RSM [28], computational intelligence capabilities are therefore essential (Figure 13.2).

In order to accurately characterize the quantitative link between the responsiveness in variables and response of the liquid flow control process, we have employed the

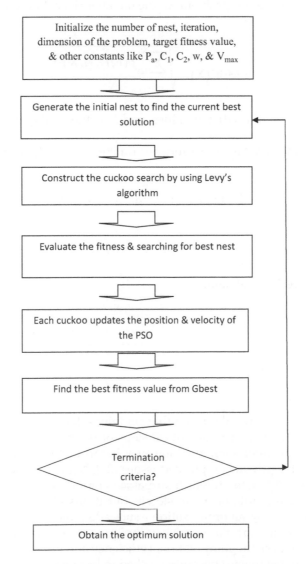

**FIGURE 13.2**
Flowchart of the proposed hybrid HCSPSO model.

well-known varying power equations ANOVA [11] and [12]. Following are some examples of how flow rate ($F$) in the mathematical model may be described in terms of the sensor output ($E$), pipe:

$$F = \mu 1.E^{\mu 2}.D^{\mu 3}.k^{\mu 4}.n^{\mu 5} \tag{13.1}$$

Where $\mu_1, \mu_2, \mu_3, \mu_4$, and $\mu_5$ are the quantitative model's parameters. Now, methods of cognitive computing are employed to extract the optimal values of this coefficient from the empirical observations dataset. As a computational intelligence tool, three distinct meta-heuristic optimization strategies can be utilized in this case to lessen the divergence between measured and simulated flow rates as well as to attain the process variable's exact efficiency and approach conformity with the observed one. RMSE was therefore employed in this study as an optimization problem for the metaheuristic that required to be whittled down.

$$\text{RMSE}(X) = \sqrt{\frac{\sum_{i=1}^{N} f(E,D,K,X)^2}{N}} \tag{13.2}$$

Where $X$ is the collection of predicted values, and $N$ is the total quantity of experimental data.

When modeling ANOVA data, the erroneous function ($E_i$, $D_i$, $k_i$, $n_i$, $X$) and set of parameters $X$ can be expressed as

$$(Ei, Di, X) = \mu_1.E^{\mu 2}.D^{\mu 3}.k^{\mu 4}.n^{\mu 5} - F \tag{13.3}$$

$$X = \{\mu_1, \mu_2, \mu_3, \mu_4, \mu_5\} \tag{13.4}$$

Where $F$ is the experimental data.

## 13.4 Proposed Methodology

### 13.4.1 Hybrid GAPSO

This section discusses a hybrid algorithm that combines the CSO and PSO strategies. The schematic in Figure 13.4 provides the critical steps of the hybrid CSPSO method. PSO is one of the most effective methodologies; however, because of how quickly it converges, it typically converges abruptly in complicated issues [29]. Numerous complicated problems are solved using the CSO method, which surpassed competing algorithms [30]. The CSO method could converge a little bit more slowly, but it has higher probing potential. No method is capable of solving all optimization issues adequately. The global optimal cause can be obtained by combining the existing techniques to handle this problem [31]. The hybrid method can enhance the computational efficiency and global optimization potential. It could make it less likely toward becoming locked in local minima. The greatest elements of both algorithms' characteristics may be combined in the hybrid algorithm.

### 13.4.2 Parameters Setting

Every calculation's configuration parameters in the test are shown as follows:

1. For CSO, extra range factor for crossover ($\gamma$) is 0.4, crossover percentage (pc) is 0.7, mutation rate (mu) is 0.1, and mutation percentage (pm) is 0.3 as indicated by the prior work [12, 32].
2. For PSO, the inertia weight damping factor ($W_{\text{damp}}$) is 0.99, inertia weight ($w$) is 1, global learning coefficient ($C_2$) is 2, and personal learning coefficient ($C_1$) is 1.5, according to the earlier work [33].
3. For HCSPSO, mutation rate is 0.05, and global learning coefficient ($C_2$) is 2, $r1$ and $r2$ are (0, 1), and degree of importance: $\alpha1$ and $\alpha2$ are (0.4, 0.4), and $\alpha3$ is 0.4 [34].

We specify 5000 maximum iterations and 100 maximum populations for each method. In order to examine optimum estimates for a liquid flow model, the search space is constrained to a 5-dimensional function optimization problem of $\{\mu_1, \mu_2, \mu_3, \mu_4, \mu_5\}$ as already shown in Equation (13.4). The search range [7] for the optimization of a liquid flow-based model is $(-15, 15)$.

Figure 13.3 shows the overall process to perform the present research. Initially, the total 134 experimental dataset was segmented into two parts: 117 data for training the model and 17 data for testing purposes. ANOVA is used as a nonlinear model to represent the present complex flow process. After obtaining the coefficient of ANOVA, testing the dataset by using the proposed three algorithms PSO, GA, and HGAPSO to get actual coefficient and optimum objective functions. Finally, the best fitted model is identified on the basis of statistical parametric criteria.

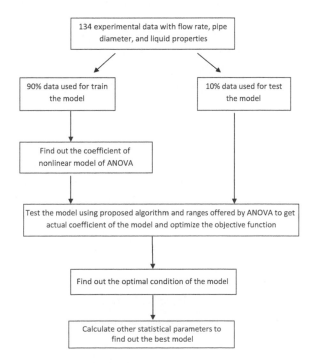

**FIGURE 13.3**
Flowchart of the overall process.

## 13.5 Performance Analysis

For the modeling of solar cells, we used Matlab 2013b version, and the specification of the PC is Intel(R) Core (TM) i3processor, 4 GB RAM with Windows 7 operating system. During these numerical experiments, we have put a strain on and thought about the proficiency of the considered calculation based on certain models such the fitness test, computational efficiency test, reliability test, convergence test, and accuracy test, which are portrayed in the following subsection separately [35]. Toward the end, overall exhibitions have been described.

### 13.5.1 Computational Efficiency Test

One of the key factors for assessing the efficacy of the bio-inspired optimization approach used in a given method of parametric optimization is computational time. For a specified number of runs of the program (5000 iterations, 100 population changes, and 10 executions of the program), we have calculated the average number of iterations for each method with each of the issues in this subsection. A study of comparability based on average execution time is presented in Table 13.1.

Table 13.1 shows that the HCSPSO model has the best average computationally expensive performance.

### 13.5.2 Convergence Speed

The conclusive outcome correlation cannot depict the searching performance of any optimization technique. An optimization technique is said to have better convergence speed when it provides the least RMSE for several runs with the variation of iteration number [36, 37]. Here we performed the convergence test by taking the iteration 100 to 5000 and each of the cases we run 20 times. All the algorithms, HCSPSO, PSO, and CSO, were run for both the models to obtain the minimum fitness. Figure 13.4 represents the convergence test for all the optimization tools. From the graph, it has been concluded that HCSPSO gives a better convergence than the other two basic metaheuristic optimization techniques.

### 13.5.3 Accuracy Test

An ANOVA method known as the accuracy test may be used to determine how closely the computed value matches the corresponding values under existing experimental settings. To quantify the discrepancy between theoretical and simulation current data, denoted as Eqs. (5) and (6), in the accuracy test, we employed two error indicator indexes: mean absolute percentage error (MAPE) and mean absolute error (MAE).

$$IAE = \left| F_{measured} - F_{calculated} \right| \tag{13.5}$$

**TABLE 13.1**

Comparative Study Based on Computational Time

| Method | Average Computational Time (Sec) |
| --- | --- |
| CSO | 159.174 |
| PSO | 119.0112 |
| HCSPSO | 21.197 |

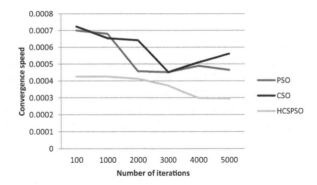

**FIGURE 13.4**
Comparative study based on convergence speed.

$$\text{MAPE} = \frac{1}{n} \sum_{i=1}^{n} \frac{IAE}{\text{Fmeasured}} \tag{13.6}$$

Moreover, MAE can be defined as:

$$\text{MAE} = \frac{\sum_{i=1}^{n} IAE_i}{n} \tag{13.7}$$

$F_{measured}$ and $F_{calculated}$ are the experimental and estimated values of liquid flow rate, respectively, where $n$ is the number of experimental datasets (Table 13.2). The best optimization algorithm consistently delivered runs with the lowest RMSE. The PSO, GA, and HCSPSO Matlab code presented in Table 13.3 is used to calculate the coefficient of the nonlinear models. The prediction error can be calculated using RMSE which can be defined as follows:

$$\text{RMSE} = \sqrt{\frac{1}{m} \sum_{i=1}^{m} \left( \frac{X_{\text{exp}} - X_{\text{Cal}}}{X_{\text{exp}}} \right)^2} *100\% \tag{13.8}$$

**TABLE 13.2**

Estimated Optimal Parameters by Using PSO, CSO, and HCSPSO Based Modeling of Liquid Flow Control Process

| Method | $\mu_1$ | $\mu_2$ | $\mu_3$ | $\mu_4$ | $\mu_5$ |
|---|---|---|---|---|---|
| PSO | 15.00 | 10.0466 | −1.0550 | −3.6296 | −0.6247 |
| CSO | 15.00 | 8.4074 | −0.7271 | −1.3846 | −0.7300 |
| HCSPSO | 9.4281 | 8.3666 | −1.2252 | 1.7282 | −1.1319 |

**TABLE 13.3**

Comparative Study Based on Mean Absolute Error (MAE) in PSO, GA, and HGAPSO

| Method | Mean Absolute Percentage Error (MAPE) | Mean Absolute Error (MAE) |
|---|---|---|
| PSO | 16.60 | 0.036 |
| CSO | 18.41 | 0.039 |
| HGAPSO | 2.84 | 0.012 |

$$\text{Accuracy} = (100 - \text{RMSE})\% \tag{13.9}$$

Where $X_{\text{exp}}$ is the experimental value, $X_{\text{cal}}$ is the calculated value, and $m$ is the number of training data. Table 13.4 describes CSO offering maximum mean absolute percentage error (MAPE) and mean absolute error (MAE). It has been also observed from Table 13.5 that HCSPSO optimization has the least RMSE error and the maximum accuracy while there is no significant difference between CSO and PSO. Figure 13.5 shows the performance metric comparison (computational time and MAPE) between CSO, PSO, and HCSPSO. Figure 13.6

**TABLE 13.4**

Comparative Study Based on Root Mean Square Error (RMSE) and Accuracy

| Method | RMSE | Accuracy |
| --- | --- | --- |
| PSO | 0.0442 | 99.9558 |
| CSO | 0.0492 | 99.9508 |
| HCSPSO | 0.0212 | 99.9788 |

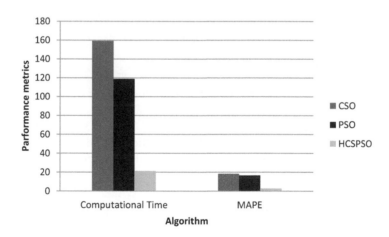

**FIGURE 13.5**
Performance metric comparison for PSO, CSO, and HCSPSO.

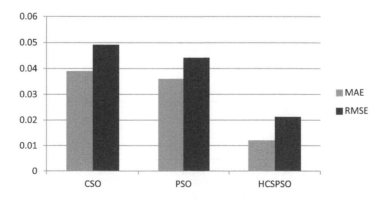

**FIGURE 13.6**
Comparison characteristics about MAE and RMSE.

shows the comparison characteristics for MAE and RMSE for all the applied algorithms in this present research. For PSO, CSO, and HCSPSO based modeling, Figure 13.7 depicts the relative errors in relation to various liquid flow rate measurement examples. Compared to previous optimization strategies, the suggested DE optimization has the lowest relative error, as can be observed.

Comparison of empirical and computed values of the outcomes are displayed in Figure 13.8 from the CSO, PSO, and HCSPSO. In comparison to practical flow rate, CSO optimization offers improved estimated flow rate. Figure 13.9 depicts the deviance graph $\left( = \dfrac{X_{exp} - X_{Cal}}{X_{exp}} \right)$ and experimental flow rate. From the graph it is seen that deviation is optimal for the entire applied algorithm during very low liquid flow, but during high flow rate, deviation is maximum for CSO but the deviation of PSO and HCSPSO is nearly saturated.

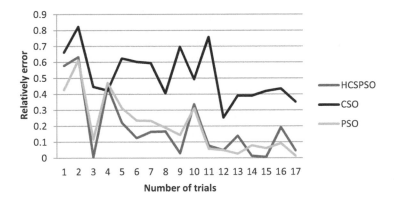

**FIGURE 13.7**
Relative errors for CSO, PSO, and HCSPSO based modeling of liquid flow control process.

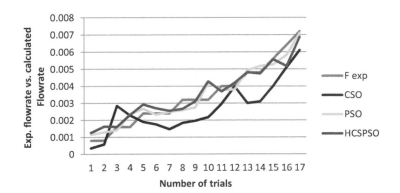

**FIGURE 13.8**
Comparisons of the characteristics of the experimental data and estimated liquid flow rate using CSO, PSO, and HCSPSO based model.

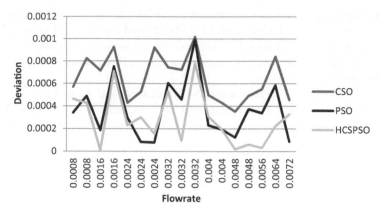

**FIGURE 13.9**
Deviation versus Experimental Flow Rate.

## 13.6 Finding Optimal Condition for Liquid Flow

The ideal pipe diameter, liquid viscosity, liquid conductivity, and sensor output voltage are searched for in the next stage of this operation in order to determine the optimum conditions for proper grinding at that moment. Now, a new fitness or objective function (OF) is defined as follows, where liquid flow rate only depends upon the potential attributes' numerical values, not their unit.

$$OF = F = \mu 1.E^{\mu 2}.D^{\mu 3}.k^{\mu 4}.n^{\mu 5}$$

$$OF = F = 9.4281.E^{8.366}.D^{-1.2252}.K^{1.7282}.n^{-1.1319} \tag{13.10}$$

In the previous section, we already get that among the applied three metaheuristic algorithms, HCSPSO is the best fitted model, which predicts the test dataset with a higher degree of accuracy. So in this section we again implemented the generated ANOVA models for liquid flow rate, which are minimized using HCSPSO. As starting solutions or demographics, a set of random values for $E$, $D$, $K$, and $n$ are utilized in this situation. Therefore, HCSPSO must optimize these four parameters in order for the value of OF to be as low as possible. These parameters' exploration ranges are chosen to align with the control factors.

After all iterations, the best conditions are discovered to be 0.218, 0.025, 0.605, and 0.898, respectively, for pipe diameter, sensor output voltage, viscosity, and liquid conductivity. This signifies that a suitable flow rate condition will occur when sensor output and conductivity is minimum, viscosity is maximum, and pipe diameter is moderate.

## 13.7 Conclusions

The research challenge of modeling and optimization of process parameters in any nonlinear process industry is intriguing. In the present research, we applied optimization techniques for the nonlinear liquid flow process model. From the experiment we obtain a total of 134 datasets among which 117 training datasets were used for designing the model while 17 test datasets were used for validation of the model. Every dataset contains the potential input variables—liquid parameters, sensor output, and pipe diameter—and the output variable of flow rate. The relationships between these input and output variables are nonlinear, so determining the ideal process parameters is genuinely challenging using conventional computing techniques.

Overall, research was performed into three stages: In the first phase, 117 datasets were used to design a nonlinear power model by ANOVA. In the second phase, two ordinary common metaheuristic optimizations, CSO and PSO, along with the improved hybrid optimization technique HCSPSO, were applied to test the model on 17 test datasets to find the coefficient of nonlinear model ANOVA. To identify the best fitted model, we used three statistical parameters—MAE, MAPE, and RMSE—and computational time. In the last phase we applied the HCSPSO algorithm for identifying the ideal values of pipe diameter, liquid viscosity, liquid conductivity, and sensor output voltage so that when liquid flow rate is reduced, the best conditions for appropriate grinding may be discovered.

Sections 13.5 and 13.6 are mainly presenting the statistical analysis of the proposed algorithm. The proposed hybrid algorithm shows better performance by means of convergence speed, computation time, and statistical error, which are the main priorities of the present research. Moreover, accuracy of the entire proposed algorithm is nearly the same and is satisfactory.

Improvement of convergence speed and computational time for modeling of any complex nonlinear model using optimization s is always an open challenge for near future. Internet of Things (IoT)–based AI models is another future scope for modeling of any complex system.

# References

1. Maia EHB, Assis LC, de Oliveira TA, da Silva AM, Taranto AG. Structure-based virtual screening: From classical to artificial intelligence. *Frontiers in Chemistry* 2020;8: 343.
2. Torun HM, Swaminathan M. High-dimensional global optimization method for high-frequency electronic design. *IEEE Transactions on Microwave Theory and Techniques* 2019;67: 2128–2142.
3. Behrooz F, Mariun N, Marhaban MH, Mohd Radzi MA, Ramli AR. Review of control techniques for HVAC systems—Nonlinearity approaches based on fuzzy cognitive maps. *Energies* 2018;11: 495.
4. Negash BM, Yaw AD. Artificial neural network based production forecasting for a hydrocarbon reservoir under water injection. *Petroleum Exploration and Development* 2020;47: 383–392.
5. Yaseen ZM, Sulaiman SO, Deo RC, Chau K-W. An enhanced extreme learning machine model for river flow forecasting: State-of-the-art, practical applications in water resource engineering area and future research direction. *Journal of Hydrology* 2019;569: 387–408.
6. Dutta P, Paul S, Obaid AJ, Pal S, Mukhopadhyay K. Feature selection based artificial intelligence techniques for the prediction of COVID like diseases. *Journal of Physics: Conference Series* 2021;1963: 012167. https://doi.org/10.1088/1742-6596/1963/1/012167
7. Dandekar R, Barbastathis G. *Neural Network aided quarantine control model estimation of COVID spread in Wuhan, China.* ArXiv:200309403 2020.
8. Dutta P, Kumar A. *Study of optimized NN model for liquid flow sensor based on different parameters.* 2018.
9. Dutta P, Kumar A. Design an intelligent flow measurement technique by optimized fuzzy logic controller. *Journal Européen Des Systèmes Automatisés* 2018;51: 89.

10. Dutta P, Kumar A. Intelligent calibration technique using optimized fuzzy logic controller for ultrasonic flow sensor. *Mathematical Modelling of Engineering Problems* 2017;4: 91–94. https://doi.org/10.18280/mmep.040205

11. Dutta P, Majumder M, Kumar A. Parametric optimization of liquid flow process by ANOVA optimized DE, PSO & GA algorithms. *International Journal of Engineering and Manufacturing (IJEM)* 2021;11: 14–24.

12. Dutta P, Kumar A. Modelling of liquid flow control system using optimized genetic algorithm. *Statistics, Optimization & Information Computing* 2020;8: 565–582.

13. Abdelsalam AM, El-Shorbagy MA. Optimization of wind turbines siting in a wind farm using genetic algorithm based local search. *Renewable Energy* 2018;123: 748–755.

14. Dutta P, Kumar A. Design an intelligent calibration technique using optimized GA-ANN for liquid flow control system. *Journal Europeen Des Systemes Automatises* 2017;50: 449–470.

15. Kumar V, Kumar A, Chhabra D, Shukla P. Improved biobleaching of mixed hardwood pulp and process optimization using novel GA-ANN and GA-ANFIS hybrid statistical tools. *Bioresource Technology* 2019;271: 274–282. https://doi.org/10.1016/j.biortech.2018.09.115

16. Liu H-G, Sun J-H, Zhang B, Zhang Y-S. ANFIS and its application in control system. *Journal of East China Shipbuilding Institute* 2001;15: 27–31.

17. Dutta P, Kumar A. Application of an anfis model to optimize the liquid flow rate of a process control system. *Chemical Engineering Transactions* 2018;71: 991–996. https://doi.org/10.3303/CET1871166

18. Dutta P, Kumar A. Modeling and optimization of a liquid flow process using an artificial neural network-based flower pollination algorithm. *Journal of Intelligent Systems* 2020;29: 787–798. https://doi.org/10.1515/jisys-2018-0206

19. Dutta P, Mandal S, Kumar A. Application of FPA and ANOVA in the optimization of liquid flow control process. *RCES* 2019;5: 7–11. https://doi.org/10.18280/rces.050102

20. Dutta P, Mandal S, Kumar A. Comparative study: FPA based response surface methodology & ANOVA for the parameter optimization in process control. *AMA_C* 2018;73: 23–27. https://doi.org/10.18280/ama_c.730104

21. Mandal S, Dutta P, Kumar A. Modeling of liquid flow control process using improved versions of elephant swarm water search algorithm. *SN Applied Sciences* 2019;1: 886. https://doi.org/10.1007/s42452-019-0914-5

22. Dutta P, Majumder M, Kumar A. An improved grey wolf optimization algorithm for liquid flow control system. *IJ Engineering and Manufacturing* 2021;4: 10–21.

23. Gálvez A, Iglesias A. A new iterative mutually coupled hybrid GA–PSO approach for curve fitting in manufacturing. *Applied Soft Computing* 2013;13: 1491–1504.

24. Jeong S, Hasegawa S, Shimoyama K, Obayashi S. Development and investigation of efficient GA/PSO-hybrid algorithm applicable to real-world design optimization. *IEEE Computational Intelligence Magazine* 2009;4: 36–44.

25. Anand A, Suganthi L. Hybrid GA-PSO optimization of artificial neural network for forecasting electricity demand. *Energies* 2018;11: 728.

26. Premalatha K, Natarajan AM. Hybrid PSO and GA for global maximization. *International Journal of Open Problems in Computer Science and Mathematics* 2009;2: 597–608.

27. Mandal S, Dutta P, Kumar A. Application of FPA and ANOVA in the optimization of liquid flow control process. *Review of Computer Engineering Studies* 2018;5: 7–11. https://doi.org/10.18280/rces.050102

28. Dutta P, Biswas SK, Biswas S, Majumder M. Parametric optimization of solar parabolic collector using metaheuristic optimization. *Computational Intelligence and Machine Learning* 2021;2: 26–32.

29. Chen M-R, Li X, Zhang X, Lu Y-Z. A novel particle swarm optimizer hybridized with extremal optimization. *Applied Soft Computing* 2010;10: 367–73. https://doi.org/10.1016/j.asoc.2009.08.014

30. Zhang M, Wang H, Cui Z, Chen J. Hybrid multi-objective cuckoo search with dynamical local search. *Memetic Computing* 2018;10: 199–208.

31. Ibrahim AM, Tawhid MA. A hybridization of cuckoo search and particle swarm optimization for solving nonlinear systems. *Evolutionary Intelligence* 2019;12: 541–61.
32. Dutta P, Majumder M. An improved grey wolf optimization technique for estimation of solar photovoltaic parameters. *International Journal of Power and Energy Systemsthis Link Is Disabled* 2021;41: 217–21.
33. Bellubbi S, Mallick B, Hameed AS, Dutta P, Sarkar MK, Nanjundaswamy S. Neural network (NN) based RSM-PSO multi-response parametric optimization of electro chemical discharge micro-machining process during micro-channel cutting on silica glass. *Journal of Advanced Manufacturing Systems* 2022;21(4): 1–29.
34. Naik R, Sathisha N. Desirability function and GA-PSO based optimization of electrochemical discharge micro-machining performances during micro-channeling on silicon-wafer using mixed electrolyte. *Silicon* 2022. https://doi.org/10.1007/s12633-022-01697-5
35. Dutta P, Cengiz K, Kumar A. Study of bio-inspired neural networks for the prediction of liquid flow in a process control system. *Cognitive Big Data Intelligence with a Metaheuristic Approach.* Elsevier;2022, pp. 173–91.
36. Dutta P, Agarwala R, Majumder M, Kumar A. Parameters extraction of a single diode solar cell model using bat algorithm, firefly algorithm & cuckoo search optimization. *Annals of the Faculty of Engineering Hunedoara* 2020;18: 147–56.
37. Dutta P, Kumar A. Design an intelligent calibration technique using optimized GA-ANN for liquid flow control system. *Journal Européen Des Systèmes Automatisés* 2017;50: 449–70. https://doi.org/10.3166/jesa.50.449-470

# 14

## Modeling of Improved Deep Learning Algorithm for Detection of Type 2 Diabetes

### 14.1 Introduction

Diabetes is a kind of disease which causes high blood sugar or glucose levels in a human body. It has three general types: in Type 1, the human body can't produce enough insulin; in Type 2, the human body can't create insulin well; and in Type 3, gestational diabetes occurs during pregnancy [1]. There are several health issues like eyes issues, kidney problems, heart problems, and strokes that occur due to diabetes. Due to this metabolic disorder, the human body neither produces or nor stores glucose for energy [2]. Hence in medical diagnosis, proper treatment of diabetes is now one of the important challenges. The number of active patients with diabetes is exponentially increased as reported by the World Health Organization (WHO). To predict the disease, different Ais have been proposed in the field of medical problems.

In artificial intelligence (AI), deep learning is one of the subparts that can self-gain from the information [3]. It is likewise fit for unaided learning by which it can get familiar with a lot of unstructured and unleveled information that even a human brain can require a long time to comprehend. Deep learning utilizes different layers like ANN to extricate the potential attributes from raw data [4]. In the previous few years, various methods have been presented; the assortment of procedures include ANN approaches and deep learning deal with analyze to diagnose diabetes.

There have been several studies performed for the diagnosis of diabetes using different feature-based deep learning algorithms; some of them are highlighted in this section. Novel deep learning (DL) approaches were studied [5] on simulated continuous glucose monitoring (CGM) signals for diagnosis of Type 2 diabetes. Among all the algorithms, CNN performs best with an average accuracy of 77.5%. A hybrid deep learning based restricted Boltzmann machine approach [6] was used to classify the states of a diabetes patient. Accuracy of the proposed model was about 92.10%. A deep neural network (DNN) was examined to predict whether a patient may have any chance of developing diabetes within five years, with the help of eight sample characteristics [7]. A number of machine learning and deep learning approaches, namely CNN, VGG-16, and VGG-19, were applied for automated classification of diabetes of retinopathy [8] by analysis of fundus images with varying illumination and fields of view. Maximum accuracy obtained by the proposed model was about 82%. A comparative study was performed using a 5-fold and 10-fold cross validation deep neural network [9] for the diagnosis of diabetes. From result analysis, it was observed that the maximum accuracy obtained by 5-fold cross validation was about 98.35%. A hybrid deep learning algorithm was modeled by a variational autoencoder (VAE), sparse

DOI: 10.1201/9781003216001-16

autoencoder (SAE), and convolutional neural network (CNN) for predicting diabetes. It was observed that the maximum accuracy obtained by the CNN classifier joined with SAE was 92.31% [10]. A comparative study was performed with three different deep learning algorithms: CNN, long short term memory (LSTM), and hybrid CNN-LSTM over a Pima Indian diabetes dataset (PIDD). Experimental results show that the proposed hybrid model CNN-LSTM predicts the classified model with highest accuracy of 91.38% [11]. A DNN was incorporated with a restricted Boltzmann machine (RBM) to analyze the Type 1 diabetes mellitus datasets after feature selection algorithm. Maximum accuracy of the proposed model is about 78% [12]. A comparison between CNN and hybrid CNN-LSTM was performed for automatically diagnosing diabetes with the help of heart rate variability (HRV) signals taken from ECG. For 5-fold cross validations, hybrid CNN-LSTM gave the maximum accuracy of about 95.1%, while CNN gave an accuracy of 93.6% [13]. An e-nose is designed for detecting the three different classes of diabetes (healthy, prediabetes, diabetes) based on the patient health data. In the final stage, an optimized DNN model was implemented for the classification of the data. The proposed systems successfully detect the multilevel diabetes with the accuracy of 96.25% [14]. A methodology for prediction of diabetes using machine learning and deep learning approaches was applied on the PIDD dataset. Among machine learning and deep learning approaches, deep learning achieved the maximum accuracy, about 98.07% [15]. A stacked auto-encoder based deep learning for the classification of Type 2 PIDD datasets contained 768 datasets with 8 attributes. The proposed algorithm achieved maximum accuracy of about 86.26% [16].

This chapter is organized as follows. Section 14.1 contains the introduction followed by some state-of-the-art techniques utilized in PIDD. Section 14.2, Methodology, contains the following subsections: description of datasets, imbalanced nature of the datasets, and how we convert this imbalanced dataset into a balanced one by using SMOTE techniques. In Sections 14.3 and 14.4, overall research flowcharts and basic information of the DNN are explained. Section 14.5 explains the result analysis, followed by the conclusion in Section 14.6.

## 14.2 Methodology

### 14.2.1 Datasets

Present research is performed on the basis of a Pima Indian diabetes dataset, which accommodates some potential attributes of the diabetes patient or the symptoms before they feel sick. The whole entity contained only female patients, none of whom have an age less than 21 years. The dataset contained a total of 768 cases; among them, 500 samples are non-diabetic and 268 samples are diabetic. Potential attributes of the Pima datasets are insulin level, age, blood pressure, skin thickness, glucose, pregnancy, and diabetes pedigree function, as shown in Table 14.1. Figure 14.1 show the box plot of each potential attribute's distribution and their variability. The PIDD is taken from the website https://data.world/data-society/pima-indians-diabetes-database.

Before applying optimization techniques, a number of potential attributes characteristic of the dataset should be identified [17]. For this purpose, the researchers applied a hyper parameters technique to identify the potential attributes. But for imbalance characteristics data, here we used the SMOTE algorithm, which is described in the following section. Cross validation is the process where experimental training output is compared with the

**TABLE 14.1**

Potential Attributes of the Dataset

| | Pregnancies | Glucose | Blood Pressure | Skin Thickness | Insulin | BMI | Diabetes Pedigree Function | Age | Outcome |
|---|---|---|---|---|---|---|---|---|---|
| 0 | 6 | 148 | 72 | 35 | 0 | 33.6 | 0.627 | 50 | 1 |
| 1 | 1 | 85 | 66 | 28 | 0 | 26.6 | 0.351 | 31 | 0 |
| 2 | 8 | 183 | 64 | 0 | 0 | 23.3 | 0.672 | 32 | 1 |
| 3 | 1 | 89 | 66 | 23 | 94 | 28.1 | 0.167 | 21 | 0 |
| 4 | 0 | 137 | 40 | 35 | 168 | 43.1 | 2.288 | 33 | 1 |

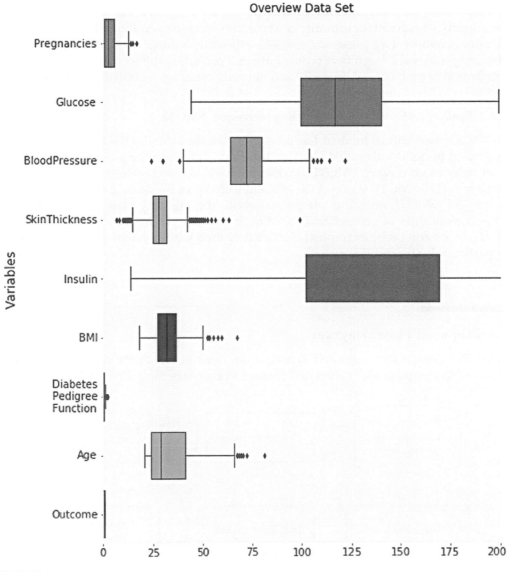

**FIGURE 14.1**

Box plots of all potential attributes.

calculated dataset output to check whether the present model is fitted best or not; after that, accuracy or other statistical metrics are calculated by applying test dataset in the fit-test model.

### 14.2.2 Imbalanced Datasets

From the PIDD dataset it is seen that out of 768 cases, positive (1) and negative (0) instances are 268 and 500, respectively. Within the dataset, there is an unbalanced distribution of classes that causes decreased classification model accuracy via unequal distribution. The fundamental reason for this is that most machine learning models are incapable of learning both positive and negative patterns. Furthermore, in this dataset, the number of positive class is less than the number of negative class, so the overall dataset is imbalanced, so that this dataset never predicts the disease with perfect accuracy. The majority of studies in the literature do not account for minority class contributions to overall classification outcomes. The uneven nature of the presented dataset is efficiently managed by SMOTE, which is one of the proposed work's significant contributions. For finding the accuracy of the model, the contribution of each dataset, majority and minority class, are recorded individually.

### 14.2.3 Synthetic Minority Over-Sampling Technique (SMOTE)

SMOTE is a well-known method for imbalance datasets [18]. For the imbalanced dataset, it was used to build a classifier. Uneven distribution of underlying output classes makes up an imbalanced dataset. SMOTE is widely utilized in the classification of datasets with imbalances [19]. SMOTE is one of the efficient techniques for dealing with an imbalanced dataset [20]. SMOTE does the interpolation with the minority class samples. This aids in the classification of generalizations. Minority classes are frequently over-sampled in SMOTE by creating false examples [21]. When dealing with unbalanced datasets, SMOTE is a particularly successful approach.

## 14.3 Proposed Flow Diagram

Figure 14.2 depicts the suggested experimental flow, in which the datasets are initially processed to eliminate null values and cleaned in preprocessing. Then SMOTE is used on

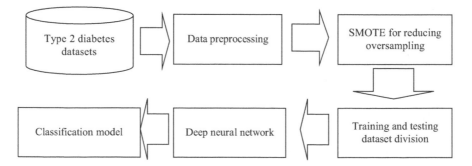

**FIGURE 14.2**
Flowcharts for the proposed model.

the given dataset to get an equal number of positive and negative samples. Furthermore, several machine learning and deep learning techniques are applied to the given dataset, and appropriate results are obtained at the end stage. For the confirmation of the results' credibility, K-Fold validation is conducted.

## 14.4  Deep Neural Network for Data Classification

In this research, an improved DNN-based framework is presented for diabetes data classification to improve the evaluation metrics of the classification problem, inspired by the intriguing properties of deep networks. The improved DNN classifier was designed by the combination of DNN and SMOTE; the model used is "Sequential"; the activation function for the DNN is chosen as "Relu" to train the inner neural network; and for the final output, the activation function used is "Softmax."

The sequential model put within the DNN is made up of two layers. There are two hidden layers in the network, each with 20 neurons. During the classification problem, the hidden layer contained a Softmax layer. The output layer will deliver the diabetic and nondiabetic class probabilities for a given record. Table 14.2 lists the parameters that were utilized to simulate the model.

## 14.5  Experimental Result Analysis

The needed machine learning algorithms are executed on the machine, which is powered by an Intel Core i3 6th generation processor, due to the modest dataset size. It contains 8 GB of RAM and a 500 GB hard drive. For diabetes data classification, a deep learning model with a Softmax layer was used. The input layer is made up of eight neurons, which correspond to the PIDD dataset's eight properties. The network is made up of two layers, each with 20 neurons that extract the data's interesting properties. To train the model and categorize the data, the Softmax layer employs a scaled conjugate gradient technique. Fine-tuning the model's parameters boosts performance by allowing the model to learn and extract information.

**TABLE 14.2**

Parameters Setting for Deep Neural Network

| Parameters | Values |
| --- | --- |
| $L_2$ weight regularization | 0.01 |
| Sparsity regularization | 5 |
| Sparsity proportion | 0.05 |
| Maximum epoch | 1000 |
| Learning rate | 0.01 |
| Loss function | Cross entropy |

### 14.5.1 Performance Measure

In the machine learning algorithm, one of the major problems is that without knowing the nature of the dataset, it generates the same type of evaluation metrics, and the best machine learning technique is identified on the basis of maximum accuracy. Machine learning techniques provide irregular accuracy and other metrics during work with an unbalanced dataset. The proposed study was evaluated using precision, recall, the F1 measure, and the receiver operating characteristics (ROC) curve [22]. Mathew's correlation coefficient (MCC) [23] is a significant parameter to judge the consistency of the model, whereas the kappa statistic is a significant metric to judge the quality of binary classifications [24]. Outcome from the present model compares with other state-of-the-art methods. In any model, the value of kappa (k) reaching unity means the model is best fitted with the experimental results; otherwise model is flawed (Tables 14.3–14.5).

### 14.5.2 Comparison with Existing System

The proposed work's results are compared to the results of other state-of-the-art existing systems in order to verify the proposed work's reliability.

Performance of different deep learning algorithms is shown in Table 14.6. Every deep learning algorithm has some drawbacks for the prediction of diabetes. Several researches have been done for training and testing of the PIDD dataset. Figure 14.3 shows the best deep learning algorithm, which gave the highest accuracies for the prediction of the PIDD dataset. Among them, the 5-fold deep neural network and hybrid CNN-LSTM offered the maximum accuracy of about 98.35% and 95.1%, respectively. But the proposed algorithm outperformed of all the state-of-the-art deep algorithms by means of highest accuracy, MCC, and K value.

**TABLE 14.3**

Average Performance Measure after Experimentation

| Algorithm | Avg. Accuracy (%) | Avg. Precision (%) | Avg. Recall (%) | Avg. F1 Score (%) | Computation Time (sec) | AUC (%) |
|---|---|---|---|---|---|---|
| SMOTE based DL | 98.60 | 98.75 | 96.2 | 97.40 | 11.25 | 98.5 |

**TABLE 14.4**

Statistical Measure after Experimentation

| Algorithm | MCC | K value |
|---|---|---|
| SMOTE based DL | 0.9288 | 0.9815 |

**TABLE 14.5**

Average Performance Measure for Positive and Negative Class after Experimentation

| Algorithm | Precision | | Recall | | F1 Score | |
|---|---|---|---|---|---|---|
| | Positive Class | Negative Class | Positive Class | Negative Class | Positive Class | Negative Class |
| SMOTE based DL | 98.5 | 99 | 94.84 | 97.56 | 96.4 | 98 |

**TABLE 14.6**

Comparative Study between Proposed Model and Existing Models

| Source | Algorithm/Technique | Maximum Accuracy |
|---|---|---|
| G. et al. (2018) | Both the CNN and CNN-LSTM algorithms were used for the detection of diabetes with the help of heart rate obtained by ECG | Best accuracy obtained by CNN-LSTM for 5-fold cross-validation was 95.1%. |
| Mohebbi et al. [5] | A novel adherence detection algorithm using LR, MLP, and CNN | Maximum accuracy obtained by CNN is 77.5%. |
| Kamble and Patil [6] | A method for diagnosing diabetes employing a DNN and 5-fold and 10-fold cross validation to train its properties | Best accuracy, F1 score, MCC obtained for 5-fold cross validation were 98.35%, 98, and 97, respectively. |
| Ashiquzzaman et al. [25] | A fully connected DNN followed by dropout layer proposed | Maximum accuracy obtained by the proposed model is 88.41%. |
| Present model | SMOTE-based deep neural network model | Maximum accuracy and MCC are 98.60% and 0.9288, respectively. |

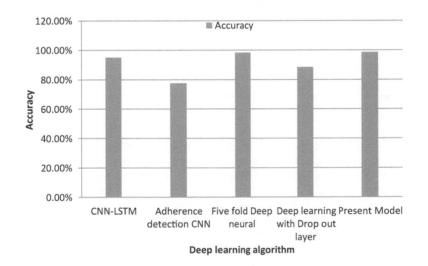

**FIGURE 14.3**
Best deep learning algorithms for the prediction of diabetes.

## 14.6 Conclusions

Proper diagnostics and disease detection in a small span of time are one of the major challenges in medical diagnosis systems. Diabetes detection is one of the major challenges in the medical domain, where delayed detection may cause the multiple organ failures including kidney failure, stroke, blindness, heart attacks, and lower limb amputation. Hence in modern research, artificial intelligence techniques were used to predict the disease accurately in a very short span of time, which helps in successful treatments. Deep learning is a subset of machine learning in AI, which has the capability of self-learning from large amounts of unstructured and unlabeled data.

As we all know, the early detection of diabetes is important as it causes other fatal diseases for human life. Many researchers have proposed different methodologies about machine learning and deep learning algorithms for the better prediction of test datasets of diabetes. Among all approaches which use the same dataset (PIDD) for their model training and testing, the deep neural network performed better. We employed 768 datasets with 17 attributes in this study, with 268 cases in the positive class (1) and 500 in the negative class (0). Within the dataset, there is an unbalanced distribution of classes. One of the key causes of decreased classification model accuracy is unequal distribution. The goal of the synthetic minority over-sampling technique (SMOTE) is to interpolate data from minority class samples in order to increase their numbers. This aids in classification generalization, and finally, a deep neural network is utilized to forecast the testing datasets. For the model validation we used the confusion matrix parameters of accuracy, recall, and F1 score and the statistical parameters of Mathew's correlation coefficient (MCC) and Kappa. In Table 14.5 we explain that the proposed SMOTE prediction is much better than the previous state-of-the-art techniques utilized for the prediction of diabetes.

Future aspects of the present research can be upgraded in the following ways: deep learning predicts accurately for image datasets, hence in the future image datasets taken by ECG, facial images, and e-nose tests of diabetes patients can be used for determining the performance of the deep learning algorithm. To improve the prediction of diabetes, IoT-based data can be used to determine the performance of deep learning by means of reduction of signal bandwidth and less computational time.

## References

1. Buchanan TA, Xiang AH, Page KA. Gestational diabetes mellitus: Risks and management during and after pregnancy. *Nature Reviews Endocrinology* 2012;8(11): 639–649.
2. Swain A, Mohanty SN, Das AC. Comparative risk analysis on prediction of diabetes mellitus using machine learning approach. *Proceedings of the 2016 International Conference on Electrical, Electronics, and Optimization Techniques (ICEEOT)*. 2016, pp. 3312–3317.
3. Lamba A, Cassey P, Segaran RR, Koh, LP. Deep learning for environmental conservation. *Current Biology* 2019;29(19): R977–R982.
4. Patel H, Thakkar A, Pandya M, Makwana K. Neural network with deep learning architectures. *Journal of Information and Optimization Sciences* 2018;39(1): 31–38.
5. Mohebbi A, Aradóttir TB, Johansen AR, Bengtsson H, Fraccaro M, Mørup M. A deep learning approach to adherence detection for type 2 diabetics. *Proceedings of the 2017 39th Annual International Conference of the IEEE Engineering in Medicine and Biology Society (EMBC)*. 2017, pp. 2896–2899. https://doi.org/10.1109/EMBC.2017.8037462
6. Kamble TP, Patil DST. Diabetes detection using deep learning approach. *International Journal for Innovative Research in Science & Technology* 2016;2(12): 8.
7. Oumaima G, Lotfi E, Fatiha E, Mohammed B. Deep learning approach as new tool for Type 2 diabetes detection. *Proceedings of the 2nd International Conference on Networking, Information Systems & Security*. 2019, pp. 1–4. https://doi.org/10.1145/3320326.3320359
8. Nguyen QH, Muthuraman R, Singh L, Sen G, Tran AC, Nguyen BP, Chua M. Diabetic retinopathy detection using deep learning. *Proceedings of the 4th International Conference on Machine Learning and Soft Computing*. 2020, pp. 103–107. https://doi.org/10.1145/3380688.3380709
9. Ayon SI, Islam M. Diabetes prediction: A deep learning approach. *International Journal of Information Engineering & Electronic Business* 2019;11(2): 21–27.
10. García-Ordás MT, Benavides C, Benítez-Andrades JA, Alaiz-Moretón H, García-Rodríguez I. Diabetes detection using deep learning techniques with oversampling and feature augmentation. *Computer Methods and Programs in Biomedicine* 2021;202: 105968. https://doi.org/10.1016/j.cmpb.2021.105968

11. Rahman M, Islam D, Mukti RJ, Saha I. A deep learning approach based on convolutional LSTM for detecting diabetes. *Computational Biology and Chemistry* 2020;88: 107329. https://doi.org/10.1016/j.compbiolchem.2020.107329

12. Balaji H, Iyenger NCSN, Caytiles R. Optimal predictive analytics of pima diabetics using deep learning. *International Journal of Database Theory and Application* 2017;10: 47–62. https://doi.org/10.14257/ijdta.2017.10.9.05

13. Swapna G, Soman K, Vinayakumar R. Automated detection of diabetes using CNN and CNN-LSTM network and heart rate signals. *Procedia Computer Science* 2018;132: 1253–1262. https://doi.org/10.1016/j.procs.2018.05.041

14. Sarno R, Sabilla SI, Wijaya DR. Electronic nose for detecting multilevel diabetes using optimized deep neural network. *Engineering Letters* 2020;28(1): 31–42.

15. Naz H, Ahuja S. Deep learning approach for diabetes prediction using Pima Indian dataset. *Journal of Diabetes & Metabolic Disorders* 2020;19(1): 391–403. https://doi.org/10.1007/s40200-020-00520-5

16. Kannadasan K, Edla DR, Kuppili V. Type 2 diabetes data classification using stacked auto-encoders in deep neural networks. *Clinical Epidemiology and Global Health* 2019;7(4): 530–535. https://doi.org/10.1016/j.cegh.2018.12.004

17. Diab S. Optimizing stochastic gradient descent in text classification based on fine-tuning hyper-parameters approach. a case study on automatic classification of global terrorist attacks. ArXiv:1902.06542; 2019.

18. Bhagat RC, Patil SS. Enhanced SMOTE algorithm for classification of imbalanced big-data using random forest. *IEEE International Advance Computing Conference (IACC)* 2015;2015: 403–408. https://doi.org/10.1109/IADCC.2015.7154739

19. Douzas G, Bacao F, Last F. Improving imbalanced learning through a heuristic oversampling method based on k-means and SMOTE. *Information Sciences* 2018;465: 1–20. https://doi.org/10.1016/j.ins.2018.06.056

20. Shuja M, Mittal S, Zaman M. Effective prediction of type II diabetes mellitus using data mining classifiers and smote. In *Advances in Computing and Intelligent Systems*, Sharma H, Govindan K, Poonia RC, Kumar S, El-Medany WM (Eds.). Springer;2020, pp. 195–211. https://doi.org/10.1007/978-981-15-0222-4_17

21. Elreedy D, Atiya AF A comprehensive analysis of synthetic minority oversampling technique (SMOTE) for handling class imbalance. *Information Sciences* 2019;505: 32–64.

22. Tripathi SL, Balas, VE, Nayak J. Comparative analysis of various supervised machine learning techniques for diagnosis of COVID-19. In *Electronic devices, circuits, and systems for biomedical applications*, Elsevier;2021, pp. 521–540.

23. Jurman G, Riccadonna S, Furlanello C. A comparison of MCC and CEN error measures in multi-class prediction. *PLoS One* 2012;7(8): e41882.

24. Wu H, Yang S, Huang Z, He J, Wang X. Type 2 diabetes mellitus prediction model based on data mining. *Informatics in Medicine Unlocked* 2018;10: 100–107.

25. Ashiquzzaman A, Tushar AK, Islam Md. R, Shon D, Im K, Park J-H, Lim D-S, Kim J. Reduction of overfitting in diabetes prediction using deep learning neural network. In *IT Convergence and Security 2017*, Kim KJ, Kim H, Baek N (Eds.). Springer;2018, pp. 35–43. https://doi.org/10.1007/978-981-10-6451-7_5

# 15

## Human Activity Recognition (HAR), Prediction, and Analysis Using Machine Learning

### 15.1 Introduction

By monitoring someone's daily behaviors, humans can learn about that person's personality and psychological state [1–3]. Following this pattern, researchers are actively researching human activity recognition (HAR), which aims to anticipate human behavior using technology. One of the crucial areas for research in computer vision and machine learning is now this. Although collecting motion data was challenging in the past, modern technical advancements make it easier for researchers to gather the data by using portable devices [4].

Machine learning (ML) is a subfield of artificial intelligence that focuses on using data to anticipate the future or assist in making decisions [5–7]. When attempting to forecast a categorical outcome, there is a categorization challenge [8, 9]. Both academics and practitioners in behavior analysis frequently base their conclusions in this field on data. These choices could involve figuring out whether an independent variable had an impact on a behavior, picking an evaluation, figuring out how conduct serves a purpose, or forecasting whether an intervention will result in significant behavior changes in a particular person. Depending upon subjectivity, the decision may vary from person to person; hence a potential solution has been made by ML [10–12]. As a result, using subjectivity to make decisions may lead to variations between behavior analysts. Intensifying the application of machine learning to behavior analysis is one possible answer to this problem. ML is widely applicable in the field of behavior experimental analysis and translational research. Additionally, some algorithms might make it easier to identify variables linked to particular behaviors that can be challenging to isolate empirically. To test hypotheses that could be challenging to evaluate with real animals, they might be simulated by machine learning [13]. In supervised ML, a model is trained using prior observations to forecast results on new samples using computerized instructions. Supervised ML algorithms have been investigated as helpful tools to enhance decision-making in a variety of sectors, including medicine, education, renewable energy, and health care, in recent years [14]. There are several research has been conducted by ML in different domains. Some of these are agriculture [15], breast cancer [16], diagnosing autism [17], detecting unsafe workplace behavior [18], renewable energy [10, 11], and so on.

One potential motivation of this research is changing the thought process of ML such that it not only addresses training or clustering the dataset but also addresses a problem

DOI: 10.1201/9781003216001-17

involving decision making in behavior analysis. The overall research work is established as follows way. In Section 15.2 previous works are discussed; the proposed methodology for HAR is presented in Section 15.3; analysis of the machine learning algorithm, result analysis, and finally conclusion and future scope of the present research are explained in Section 15.4–15.6, respectively.

## 15.2 Related Works

There are several research works that have been done in the past; some of these are highlighted here. A smart home sensor has been used to collect the data, and finally a long short term memory (LSTM) model is used to analyze the data for human activity recognition followed by data acquisition in ANN [19, 20]. A smartphone sensor based physical activity monitoring device has been proposed [21–27] to monitor the physical activities such as walking, jogging, climbing, cycling, driving, and so on. To determine its accuracy, four different ML classifier are used: naïve Bayes, decision trees, k-nearest neighbor (KNN), and support vector machine (SVM). They attain a true positive rate of more than 95% and a false positive rate of less than 1.5%. A robust HAR system has been implemented with the help of coordinate transformation, PCA, and online SVM [28, 29]. To learn the impact of orientation fluctuations, coordinate transformation and principal component analysis (CT-PCA) has been utilized. Their proposed one class support vector machine (OSVM) is independent and only makes use of a little amount of information from the hidden location. Through smartphone accelerometer data, [30, 31] offer a HAR method with variable location and orientation. In the model, authors additionally incorporate both generic and site-specific SVM. Many researchers primarily use general machine learning techniques since deep learning (DL) approaches need a significant number of datasets [32].

## 15.3 Proposed Method for Human Action Recognition

### 15.3.1 Data Collection Overview

In this research, the experimental dataset has been acquired from 30 people in the age range of 19 to 48 years who participated in the trials. Six fundamental tasks were carried out by them: three static postures and three dynamic activities [31, 32]. Between the static postures in the trial, postural changes also took place: stand to sit, sit to stand, lie to sit, stand to lie, and lie to stand. We recorded 3-axial linear acceleration and 3-axial angular velocity at a constant rate of 50 Hz using the device's built-in accelerometer and gyroscope.

### 15.3.2 Signal Processing

After applying noise filters as a preprocessing step, the accelerometer and gyroscope sensor data were sampled with fixed-width sliding windows of 2.56 seconds. A Butterworth low-pass filter was used to separate the gravitational and body motion component attributes from the actual datasets [33].

### 15.3.3 Feature Selection

The features selected for this database come from the accelerometer and gyroscope 3-axial raw signals tAcc-XYZ and tGyro-XYZ. These time domain signals were captured at a constant rate of 50 Hz. To eliminate the noise of the body acceleration signal, a 3rd-order Butterworth low pass filter with a corner frequency of 20 Hz was used, and for gravity acceleration signals Butterworth low pass filter with a corner frequency of 0.3 Hz was utilized.

### 15.3.4 Exploratory Data Analysis

The University of California Irvine (UCI) machine learning repository is where the dataset was taken from [34]. The data collection included 30 volunteers who were between the ages of 19 and 48. Each participant completed six tasks described in a previous subsection. A total of 561 feature vectors with time- and frequency-domain variables make up the dataset [35–37]. There are 10,299 records in total, split 70/30 between the training and test datasets. The six classes are then translated to numbers in the following order: [1–6].

### 15.3.5 Data Preprocessing

Duplicates and the missing values were checked, and no duplicates or missing values were found [38–40]. Next, the data imbalance was checked, the output of which is depicted in Figure 15.2. Various human activities and their counts are included in the training dataset. Thus, we discovered that there are about the same number of observations across all six activities, proving that there is no class imbalance in the data.

### 15.3.6 Exploratory Data Analysis for Static and Dynamic Activities

Static activities: no moving activities are shown in Figure 15.3.

From Figure 15.3, static and dynamic activities can be separated.

```
if(tBodyAccMag-mean()<=-0.5):
    Activity = "static"
else:
    Activity = "dynamic"
```

Figure 15.4 presents the graphical representation of static and dynamic activities. Figure 15.5 presents a box plot of body acceleration magnitude mean across all the six categories.

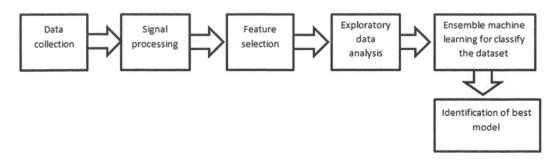

**FIGURE 15.1**
Proposed models for HAR in present research.

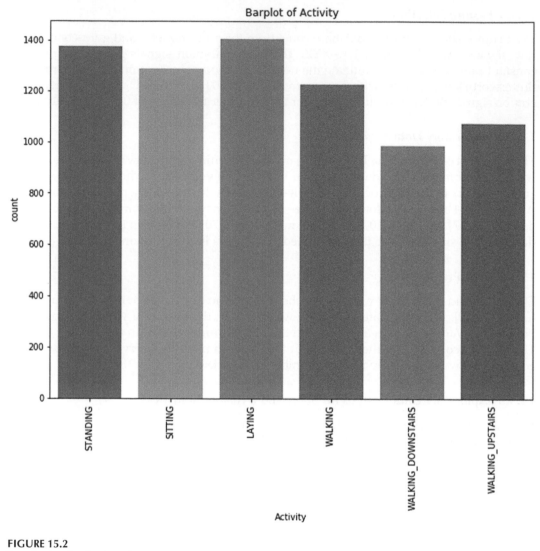

**FIGURE 15.2**
Bar plot for distribution of activities.

```
if(tBodyAccMag-mean()<=-0.8):
    Activity = "static"
if(tBodyAccMag-mean()>=-0.6):
    Activity = "dynamic"
```

Also, we can easily separate the WALKING_DOWNSTAIRS activity from others using a box plot.

```
if(tBodyAccMag-mean()>0.02):
    Activity = "WALKING_DOWNSTAIRS"
else:
    Activity = "others"
```

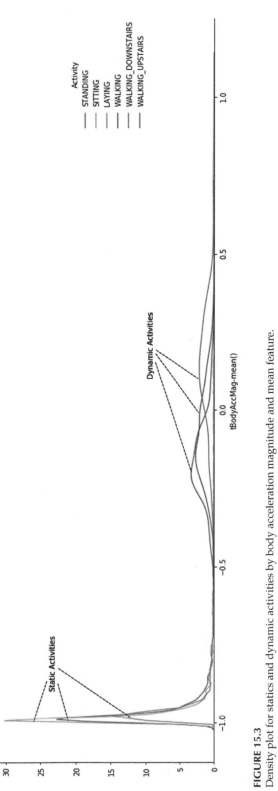

**FIGURE 15.3**
Density plot for statics and dynamic activities by body acceleration magnitude and mean feature.

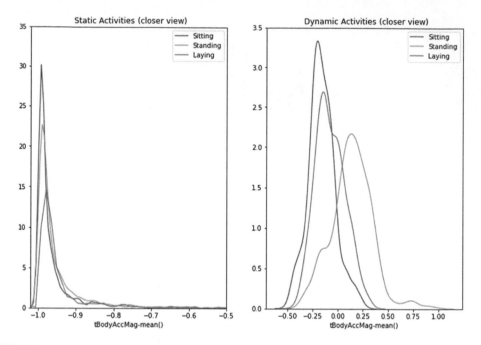

**FIGURE 15.4**
Graphical representation of static and dynamic activities.

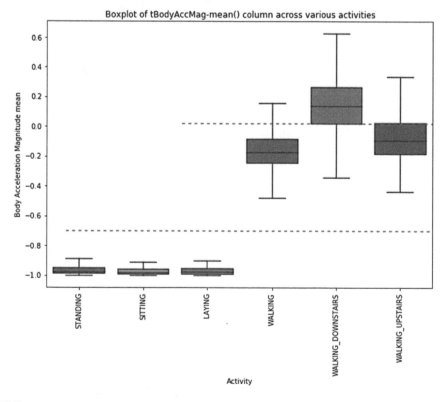

**FIGURE 15.5**
Box plot for body acceleration—magnitude mean characteristics.

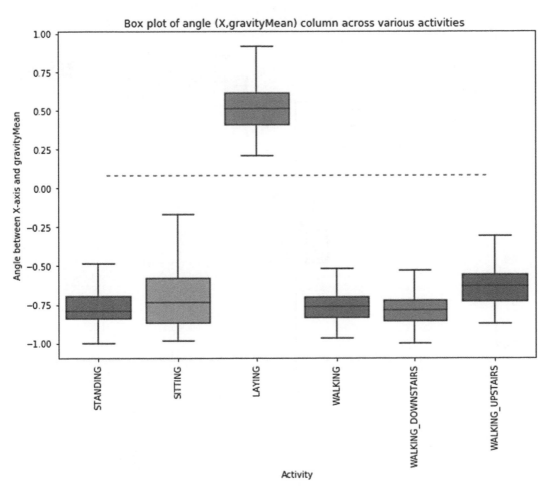

**FIGURE 15.6**
Analyzing angle between *x*-axis and gravity mean feature.

Angle(X,gravityMean) perfectly separates LAYING from other activities as shown in Figure 15.6.

```
if(angle(X,gravityMean)>0.01):
    Activity = "LAYING"
else:
    Activity = "others"
```

Similarly, the angle between the *y* axes separated from gravity feature is shown in Figure 15.7.

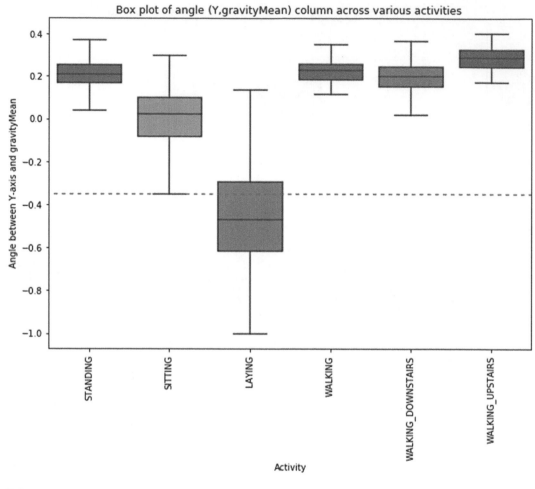

**FIGURE 15.7**
Analyzing angle between *y*-axis and gravity mean feature.

### 15.3.7 Visualizing Data Using t-SNE

A very good visual separation can be achieved in two dimensions using the PCA method [41]. Hence an experiment with multivariate dimensionality reduction can be done efficiently. The display of high-dimensional datasets is especially well suited for the nonlinear dimensionality reduction method known as t-distributed stochastic neighbor embedding (t-SNE) [42]. It is widely used in the processing of voice, genetic data, natural language processing (NLP), and image data. It is possible to display t-SNE data from an extremely high dimensional space to a low dimensional environment while still maintaining a significant amount of real information [43, 44]. Through t-SNE, each of the six activities in a 2D space, given that the training data comprises 561 distinct features, can viewed and differentiable (Figure 15.8).

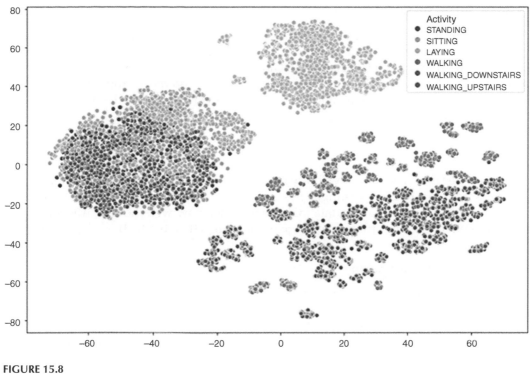

**FIGURE 15.8**
t-SNE activity visualization.

## 15.4 Machine Learning Algorithm

### 15.4.1 Logistics Regression

As a benchmark model, logistic regression (LR) always optimizes the probability of the data. This model is simple and lacks the intricacy of neural networks. If the data is on the proper side of the separating hyper plane, LR performs better. To compare the performance, a kernel technique with "lbfgs" optimizer can be used. With that mix of LR, some prior study has produced some promising outcomes [44–46]. Due to its variant this algorithm is suitable for the present research (Table 15.1).

**TABLE 15.1**

Best Parameters for LR Model [47]

| Parameters | Values | Description |
| --- | --- | --- |
| Regularizer | L2 [L1, L2] | Regularization |
| C | 2 [0.1, 0.5, 1, 2, 5, 10, 100] | Penalty parameter C |

### 15.4.2 Random Forest

An amalgam of several decision trees makes up the random forest (RF) model. It integrates many teaching techniques and finally improves the overall result. Here, the criterion assessed is nothing but the level of a split. Gini is used to check for impurities, and entropy provides the information. The parameter is N estimators, and the maximum depth is used. The number of trees in the created forest is N estimators [48, 49], and the maximum depth of a tree is max depth (Table 15.2).

### 15.4.3 Decision Tree

A decision tree is a choice and possibility formation that resembles a tree [51]. By grouping instances from root to leaf, it distinguishes between them. Here, a Gini is used to determine impurity and entropy to gather data [52]. These gauge how well the split was made. The highest depth of the tree utilizes the range from 0 to 8 [53] (Table 15.3).

### 15.4.4 Support Vector Machine

One of the selected ML algorithms is the support vector machine (SVM) [54]. A separating hyper plane determines the discriminative classifier SVM. A new data point can be categorized by the SVM optimum hyper plane [55]. Each class is located on either side of this hyper plane in two-dimensional space [56]. The support vector classifier (SVC) for more

**TABLE 15.2**

Best Parameters for RF Model [50]

| Parameters | Values | Description |
|---|---|---|
| Criterion | Gini [Gini, Entropy] | Gini impurity and entropy is information gain based |
| N Estimators | 90 [10, 150] | Trees number |
| Max depth | 6 [0,2, 4, 6, 8, 10] | Highest depth of the tree |

**TABLE 15.3**

Best Parameters for DT Model [52]

| Parameters | Values | Description |
|---|---|---|
| Criterion | Gini [Gini, Entropy] | Gini impurity and Entropy is information gain based |
| Max depth | 4 [0, 2, 4, 6, 8, 10] | Highest depth of the tree |

**TABLE 15.4**

Best Parameters for SVM Model [56]

| Parameters | Values | Description |
|---|---|---|
| Kernel | Linear [Linear, RBF, Sigmoid] | Kernel type model used |
| C | 1 [0.1, 0.5, 1, 2, 5, 10, 100] | C is a penalty parameter |

than two classes is one of the SVM variations. In this study, SVC with kernels from the "linear," "rbf," and "sigmoid" families is used. When there are many features, SVM often offers decent accuracy (Table 15.4).

### 15.4.5 K Nearest Neighbor (KNN)

KNN is one of the simplest nonparametric supervised ML algorithms [57, 58]. This algorithm is mainly used to identify in which specific category an unknown data points lies using the concept of Euclidean distance and specific value of $K$. The classification of a dataset can be identify based on the value of $K$ ranging from 1 to 5, defaulter metric, Minkowski, and standard Euclidean metric ($p$) [59] (Table 15.5).

### 15.4.6 Naïve Bayes

NB is a kind of probability-based ML technique which classifies the objects on certain features [60–62]. In NB, features are independent. One particular feature is not sufficient to discriminate one object from another. Hence it is called naïve. There are three different types of naïve classifier model used: multinomial, Bernoulli, and Gaussian (Table 15.6).

### 15.4.7 Data Preprocessing

There are train and test parts of the dataset. There are no redundant or missing values. The data frame contains both train and test data at first. To shrink the feature space, we need to delete some features. By computing for the dataset, we attempted to choose features. With the highest ANOVA F-values, the top 100, 200, 400, and 500 attributes were considered. The accuracy did, however, dramatically improve each time after adding the characteristics for all five algorithms, as seen in Figure 15.9. As a result, we test and train using all the functionalities. There are two sections to the learning. All of the algorithms' ideal parameters are initially learned. The algorithms are then contrasted in the following phase. Here $k = 5$ is used to evaluate the cross-validation dataset. Stratified sampling is used to divide the data. Without modifying the test data, we cross-validate using 70% of the training data.

**TABLE 15.5**

Best Parameters for KNN Model

| Parameters | Values | Description |
| --- | --- | --- |
| N_neighbors | 5 [1,5] | Defines the required neighbors of the algorithm |
| Metrics | Minkowski | Decides the distance between the points. |
| p | 2 | Standard Euclidean metric |

**TABLE 15.6**

Best Parameters for Naïve Bayes Model

| Parameters | Values | Description |
| --- | --- | --- |
| var_smoothing | default = 1e-9 | Variances for calculation stability |

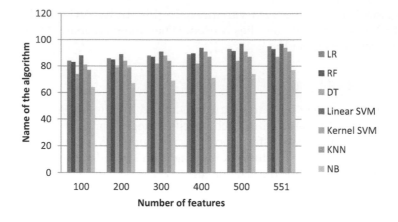

**FIGURE 15.9**
Accuracy of the seven algorithms by increasing features.

## 15.5 Experimental Results

To identify the ideal hyper parameters, here a grid search and 5-fold stratified cross-validation is used. We run the best parameter models on the test data; after that, a statistical significance test is measured. In evaluation metrics, accuracy is calculated by comparing total number of forecasts to the number of right predictions [63–65]. This parameter is essential because it describes correctly whether a person is sitting or walking. The main metric used to differentiate between these models is accuracy. Using true positive (TP), true negative (TN), false positive (FP), and false negative (FN) in Equation (15.1), we obtain an accuracy score.

$$\text{Accuracy} = \frac{\text{TP} + \text{TN}}{\text{TP} + \text{TN} + \text{FP} + \text{FN}} \tag{15.1}$$

Additionally, we use precision, recall, and F1 score as the evaluation metrics. The equation of these three metrics is given in Equations (15.2)–(15.4).

$$\text{Precision} = \frac{\text{TP}}{\text{TP} + \text{FP}} \tag{15.2}$$

$$\text{Recall} = \frac{\text{TP}}{\text{TP} + \text{FN}} \tag{15.3}$$

$$\text{F1} = \frac{2 * \text{Precision} * \text{Recall}}{\text{Precision} + \text{Recall}} \tag{15.4}$$

Figures 15.10 and 15.11 show the confusion matrix and classification report for logistics regression; Figures 15.12 and 15.13 show random forest; Figures 15.14 and 15.15 show decision trees; Figures 15.16 and 15.17 show linear SVM; Figures 15.18 and 15.19 show kernel SVM; Figures 15.20 and 15.21 show KNN; and Figures 15.22 and 15.23 show naïve Bayes.

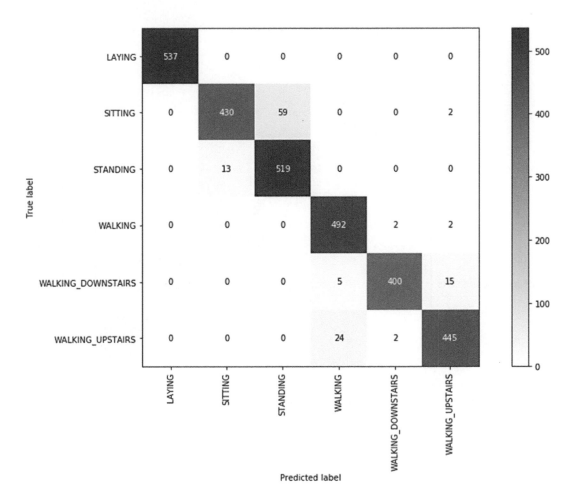

**FIGURE 15.10**
Confusion matrix for LR.

```
Classification Report
                     precision    recall   f1-score   support

            LAYING       1.00       1.00      1.00       537
           SITTING       0.97       0.88      0.92       491
          STANDING       0.90       0.98      0.94       532
           WALKING       0.94       0.99      0.97       496
WALKING_DOWNSTAIRS       0.99       0.95      0.97       420
  WALKING_UPSTAIRS       0.96       0.94      0.95       471

          accuracy                            0.96      2947
         macro avg       0.96       0.96      0.96      2947
      weighted avg       0.96       0.96      0.96      2947

accuracy:

0.9579233118425518
```

**FIGURE 15.11**
Classification report for LR.

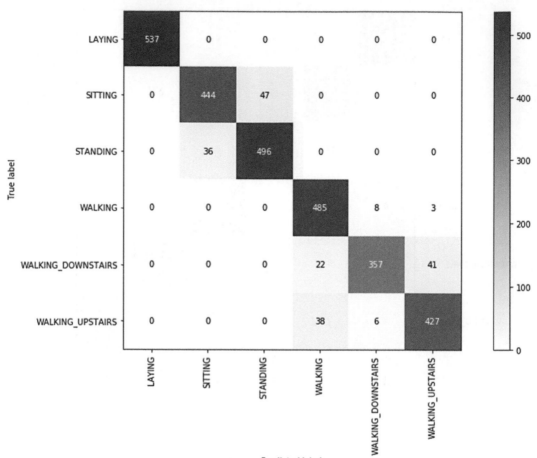

**FIGURE 15.12**
Confusion matrix for RF.

```
Classification Report
                     precision    recall  f1-score   support

            LAYING       1.00      1.00      1.00       537
           SITTING       0.93      0.90      0.91       491
          STANDING       0.91      0.93      0.92       532
           WALKING       0.89      0.98      0.93       496
WALKING_DOWNSTAIRS       0.96      0.85      0.90       420
  WALKING_UPSTAIRS       0.91      0.91      0.91       471

          accuracy                          0.93      2947
         macro avg       0.93      0.93      0.93      2947
      weighted avg       0.93      0.93      0.93      2947

accuracy:

0.9317950458092976
```

**FIGURE 15.13**
Classification report for RF.

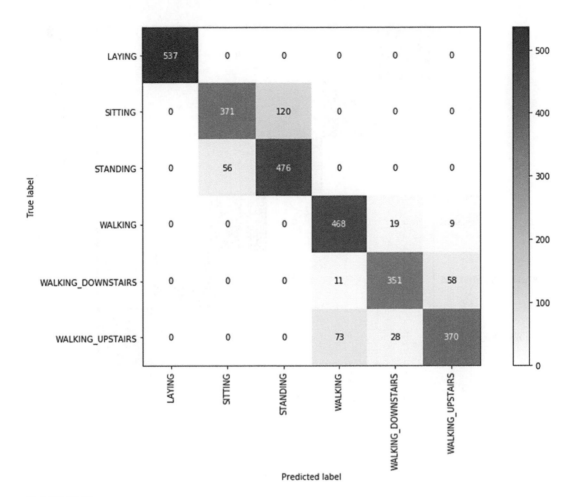

**FIGURE 15.14**
Confusion matrix for DT.

```
Classification Report
                    precision    recall  f1-score   support

            LAYING       1.00      1.00      1.00       537
           SITTING       0.87      0.76      0.81       491
          STANDING       0.80      0.89      0.84       532
           WALKING       0.85      0.94      0.89       496
WALKING_DOWNSTAIRS       0.88      0.84      0.86       420
  WALKING_UPSTAIRS       0.85      0.79      0.81       471

          accuracy                          0.87      2947
         macro avg       0.87      0.87      0.87      2947
      weighted avg       0.87      0.87      0.87      2947

accuracy:

0.8730912792670512
```

**FIGURE 15.15**
Classification report for DT.

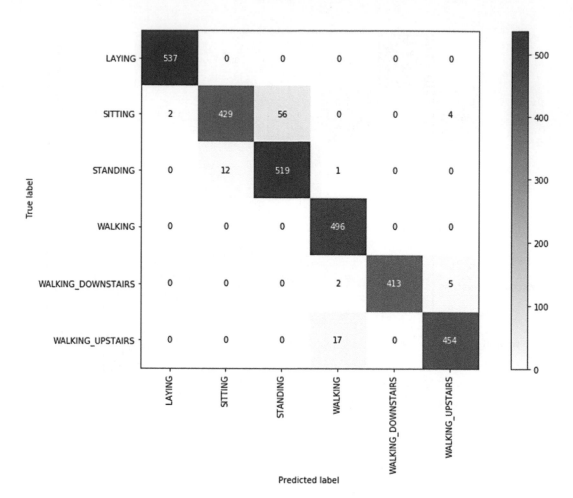

**FIGURE 15.16**
Confusion matrix for linear SVM.

```
Classification Report
                    precision    recall  f1-score   support

            LAYING       1.00      1.00      1.00       537
           SITTING       0.97      0.87      0.92       491
          STANDING       0.90      0.98      0.94       532
           WALKING       0.96      1.00      0.98       496
WALKING_DOWNSTAIRS       1.00      0.98      0.99       420
  WALKING_UPSTAIRS       0.98      0.96      0.97       471

          accuracy                          0.97      2947
         macro avg       0.97      0.97      0.97      2947
      weighted avg       0.97      0.97      0.97      2947

accuracy:

0.9664065151001018
```

**FIGURE 15.17**
Classification report for linear SVM.

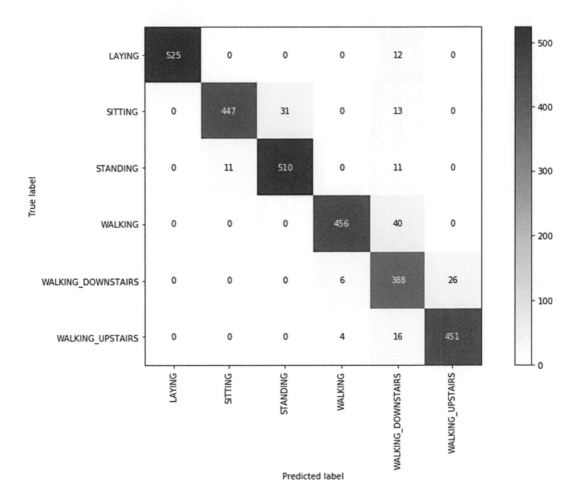

**FIGURE 15.18**
Confusion matrix for kernel SVM.

```
Classification Report
                    precision    recall  f1-score   support

            LAYING       1.00      0.98      0.99       537
           SITTING       0.98      0.91      0.94       491
          STANDING       0.94      0.96      0.95       532
           WALKING       0.98      0.92      0.95       496
WALKING_DOWNSTAIRS       0.81      0.92      0.86       420
  WALKING_UPSTAIRS       0.95      0.96      0.95       471

          accuracy                          0.94      2947
         macro avg       0.94      0.94      0.94      2947
      weighted avg       0.95      0.94      0.94      2947

accuracy:

: 0.9423142178486597
```

**FIGURE 15.19**
Classification report for kernel SVM.

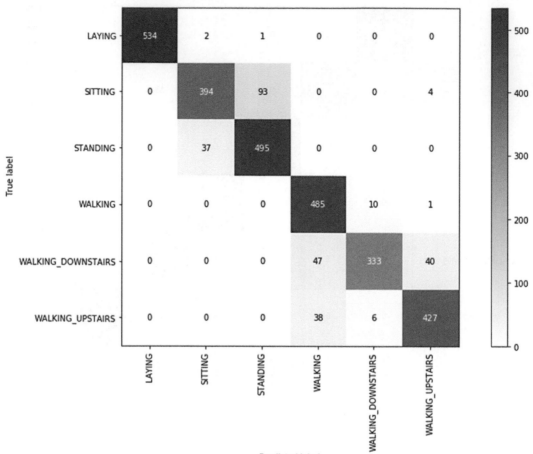

**FIGURE 15.20**
Confusion matrix for KNN.

```
Classification Report
                    precision    recall  f1-score   support

            LAYING       1.00      0.99      1.00       537
           SITTING       0.91      0.80      0.85       491
          STANDING       0.84      0.93      0.88       532
           WALKING       0.85      0.98      0.91       496
WALKING_DOWNSTAIRS       0.95      0.79      0.87       420
  WALKING_UPSTAIRS       0.90      0.91      0.91       471

          accuracy                          0.91      2947
         macro avg       0.91      0.90      0.90      2947
      weighted avg       0.91      0.91      0.90      2947

accuracy:

0.9053274516457415
```

**FIGURE 15.21**
Classification report for KNN.

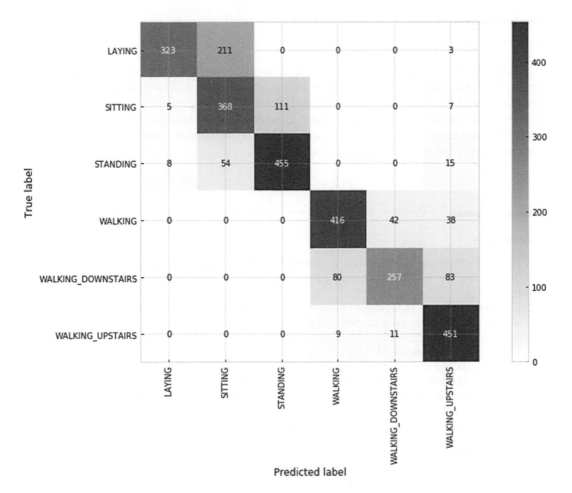

**FIGURE 15.22**
Confusion matrix for naïve Bayes.

```
          Classification Report
                          precision    recall  f1-score   support

                 LAYING       0.96      0.60      0.74       537
                SITTING       0.58      0.75      0.65       491
               STANDING       0.80      0.86      0.83       532
                WALKING       0.82      0.84      0.83       496
     WALKING_DOWNSTAIRS       0.83      0.61      0.70       420
       WALKING_UPSTAIRS       0.76      0.96      0.84       471

               accuracy                           0.77      2947
              macro avg       0.79      0.77      0.77      2947
           weighted avg       0.79      0.77      0.77      2947

          accuracy:

          0.7702748557855447
```

**FIGURE 15.23**
Classification report for naïve Bayes.

## 15.6 Conclusion

In conclusion, it is evident that linear SVM performs better than the other six algorithms, though RF, kernel VM, and LR are closer in case of accuracy. Naïve Bayes cannot compete with linear and kernel SVM for this dataset. To explore the performance of human activity recognition, a hybrid machine learning algorithm can be used as a future scope of the research. In addition, a significant amount of hyper parameter tuning is required before other machine learning techniques are applied.

## References

1. Popoola OP, Wang K. Video-based abnormal human behavior recognition—A review. *IEEE Transactions on Systems, Man, and Cybernetics, Part C (Applications and Reviews)* 2012;42(6): 865–878.
2. Candamo J, Shreve M, Goldgof DB, Sapper DB, Kasturi R. Understanding transit scenes: A survey on human behavior-recognition algorithms. *IEEE Transactions on Intelligent Transportation Systems* 2009;11(1): 206–224.
3. Rodríguez ND, Cuéllar MP, Lilius J, Calvo-Flores MD. A survey on ontologies for human behavior recognition. *ACM Computing Surveys (CSUR)* 2014;46(4): 1–33.
4. Bruno B, Mastrogiovanni F, Sgorbissa A, Vernazza T, Zaccaria R. Analysis of human behavior recognition algorithms based on acceleration data. *Proceedings of the 2013 IEEE International Conference on Robotics and Automation.* IEEE;2013, pp. 1602–1607.
5. Dutta P, Paul S, Kumar A. Comparative analysis of various supervised machine learning techniques for diagnosis of COVID-19. In *Electronic devices, circuits, and systems for biomedical applications*, Tripathi SL, Balas VE, Nayak J (Eds.). Academic Press;2021, pp. 521–540.
6. Dutta P, Paul S, Obaid AJ, Pal S, Mukhopadhyay K. Feature selection based artificial intelligence techniques for the prediction of COVID like diseases. *Journal of Physics: Conference Series* 2021, July:1963(1): 012167.
7. Dutta P, Paul S, Shaw N, Sen S, Majumder M. Heart disease prediction: A comparative study based on a machine-learning approach. In *Artificial Intelligence and Cybersecurity*, Priyadarshini I, Sharma R (Eds.). CRC Press;2022, pp. 1–18.
8. Dutta P, Paul S, Majumder M. *Intelligent SMOTE Based Machine learning Classification for Fetal State on Cardiotocography Dataset.* Researchsquare;2021.
9. Dutta P, Paul S, Majumder M. *An efficient SMOTE based machine learning classification for prediction & detection of PCOS.* Researchsquare;2021.
10. Sen S, Saha S, Chaki S, Saha P, Dutta P. Analysis of PCA based AdaBoost machine learning model for predict mid-term weather forecasting. *Computational Intelligence and Machine Learning* 2021;2(2): 41–52.
11. Duttaa P, Shawb N, Dasc K, Ghosh L. Early & accurate forecasting of mid term wind energy based on PCA empowered supervised regression model. *Computational Intelligence and Machine Learning* 2021;2(1): 53–64.
12. Greener JG, Kandathil SM, Moffat L, Jones DT. A guide to machine learning for biologists. *Nature Reviews Molecular Cell Biology* 2022;23(1): 40–55.
13. Carleo G, Cirac I, Cranmer K, Daudet L, Schuld M, Tishby N, … Zdeborová L. Machine learning and the physical sciences. *Reviews of Modern Physics* 2019;91(4): 045002.
14. Sun S, Cao Z, Zhu H, Zhao J.A survey of optimization methods from a machine learning perspective. *IEEE Transactions on Cybernetics* 2019;50(8): 3668–3681.

15. Liakos KG, Busato P, Moshou D, Pearson S, Bochtis D. Machine learning in agriculture: A review. *Sensors* 2018;18(8): 2674.
16. Bayrak EA, Kırcı P, Ensari T. Comparison of machine learning methods for breast cancer diagnosis. *Proceedings of the 2019 Scientific Meeting on Electrical-Electronics & Biomedical Engineering and Computer Science (EBBT)*, IEEE;2019, pp. 1–3.
17. Nogay HS, Adeli H. Machine learning (ML) for the diagnosis of autism spectrum disorder (ASD) using brain imaging. *Reviews in the Neurosciences* 2020;31(8): 825–841.
18. Pishgar M, Issa SF, Sietsema M, Pratap P, Darabi H. REDECA: A novel framework to review artificial intelligence and its applications in occupational safety and health. *International Journal of Environmental Research and Public Health* 2021;18(13): 6705.
19. Oluwalade B, Neela S, Wawira J, Adejumo T, Purkayastha S. Human activity recognition using deep learning models on smartphones and smartwatches sensor data. arXiv:2103.03836;2021.
20. Ramasamy Ramamurthy S, Roy N. Recent trends in machine learning for human activity recognition—A survey. *Wiley Interdisciplinary Reviews: Data Mining and Knowledge Discovery* 2018;8(4): e1254.
21. Demrozi F, Pravadelli G, Bihorac A, Rashidi P. Human activity recognition using inertial, physiological and environmental sensors: A comprehensive survey. *IEEE Access* 2020;8: 210816–210836.
22. Chelli A, Pätzold M. A machine learning approach for fall detection and daily living activity recognition. *IEEE Access* 2019;7: 38670–38687.
23. Hofmann C, Patschkowski C, Haefner B, Lanza G. Machine learning based activity recognition to identify wasteful activities in production. *Procedia Manufacturing* 2020;45: 171–176.
24. Jobanputra C, Bavishi J, Doshi N. Human activity recognition: A survey. *Procedia Computer Science* 2019;155: 698–703.
25. Jmal A, Barioul R, Meddeb Makhlouf A, Fakhfakh A, Kanoun O. An embedded ANN raspberry PI for inertial sensor based human activity recognition. *International Conference on Smart Homes and Health Telematics*. Springer, Cham;2020, pp. 375–385.
26. Voicu RA, Dobre C, Bajenaru L, Ciobanu RI. Human physical activity recognition using smartphone sensors. *Sensors* 2019;19(3): 458.
27. Qi J, Yang P, Waraich A, Deng Z, Zhao Y, Yang Y. Examining sensor-based physical activity recognition and monitoring for healthcare using Internet of Things: A systematic review. *Journal of Biomedical Informatics* 2018;87: 138–153.
28. Qi W, Wang N, Su H, Aliverti A. DCNN based human activity recognition framework with depth vision guiding. *Neurocomputing* 2022;486: 261–271.
29. Kumar M, Gautam P, Semwal VB. Dimensionality reduction-based discriminatory classification of human activity recognition using machine learning. *Proceedings of 3rd International Conference on Computing, Communications, and Cyber-Security*. Springer, Singapore;2023, pp. 581–593.
30. Hassan MM, Uddin MZ, Mohamed A, Almogren A. A robust human activity recognition system using smartphone sensors and deep learning. *Future Generation Computer Systems* 2018;81: 307–313.
31. Yang Y, Hou C, Lang Y, Guan D, Huang D, Xu J. Open-set human activity recognition based on micro-Doppler signatures. *Pattern Recognition* 2019;85: 60–69.
32. Zhou X, Liang W, Kevin I, Wang K, Wang H, Yang LT, Jin Q. Deep-learning-enhanced human activity recognition for Internet of healthcare things. *IEEE Internet of Things Journal* 2020;7(7): 6429–6438.
33. Chen K, Zhang D, Yao L, Guo B, Yu Z, Liu Y. Deep learning for sensor-based human activity recognition: Overview, challenges, and opportunities. *ACM Computing Surveys (CSUR)* 2021;54(4): 1–40.
34. Milo T, Somech A. Automating exploratory data analysis via machine learning: An overview. *Proceedings of the 2020 ACM SIGMOD International Conference on Management of Data*. 2020, June, pp. 2617–2622.
35. Majumder MG, Gupta SD, Paul J Perceived usefulness of online customer reviews: A review mining approach using machine learning & exploratory data analysis. *Journal of Business Research* 2022;150: 147–164.

36. Shabbir A, Shabbir M, Javed AR, Rizwan M, Iwendi C, Chakraborty C. Exploratory data analysis, classification, comparative analysis, case severity detection, and internet of things in COVID-19 telemonitoring for smart hospitals. *Journal of Experimental & Theoretical Artificial Intelligence* 2022;1–28. https://doi.org/10.1080/0952813X.2021.1960634

37. Taboada GL, Han L. Exploratory data analysis and data envelopment analysis of urban rail transit. *Electronics* 2020;9(8): 1270.

38. Mishra P, Biancolillo A, Roger JM, Marini F, Rutledge DN. New data preprocessing trends based on ensemble of multiple preprocessing techniques. *TrAC Trends in Analytical Chemistry* 2020;132: 116045.

39. Renner G, Nellessen A, Schwiers A, Wenzel M, Schmidt TC, Schram J. Data preprocessing & evaluation used in the microplastics identification process: A critical review & practical guide. *TrAC Trends in Analytical Chemistry* 2019;111: 229–238.

40. Baek S, Lee I. Single-cell ATAC sequencing analysis: From data preprocessing to hypothesis generation. *Computational and Structural Biotechnology Journal* 2020;18: 1429–1439.

41. Devassy BM, George S. Dimensionality reduction and visualisation of hyperspectral ink data using t-SNE. *Forensic Science International* 2020;311: 110194.

42. Hajibabaee P, Pourkamali-Anaraki F, Hariri-Ardebili MA. An empirical evaluation of the t-SNE algorithm for data visualization in structural engineering. *Proceedings of the 2021 20th IEEE International Conference on Machine Learning and Applications (ICMLA)*. IEEE;2021, December, pp. 1674–1680.

43. Zhou B, Jin W. Visualization of single cell RNA-Seq data using t-SNE in R. In *Stem Cell Transcriptional Networks*, John M. Walker (Eds.). Humana;2020, pp. 159–167.

44. Melit Devassy B, George S, Nussbaum P. Unsupervised clustering of hyperspectral paper data using t-SNE. *Journal of Imaging* 2020;6(5): 29.

45. Bukhari MM, Ghazal TM, Abbas S, Khan MA, Farooq U, Wahbah H, … Adnan KM. An intelligent proposed model for task offloading in fog-cloud collaboration using logistics regression. *Computational Intelligence and Neuroscience* 2022;2022: 1–25.

46. Adenomon MO, John DO. *Modelling hypertension and risk factors among adults using ordinal logistics regression model.* preprints, 2020, pp. 1–23. https://doi.org/10.20944/preprints202001.0291.v1

47. Crisostomo ASI, Al Dhuhli BS, Gustilo RC. Situational analysis of remote learning using logistics regression and decision trees amidst Covid19 pandemic. *Proceedings of the 2021 5th World Conference on Smart Trends in Systems Security and Sustainability (WorldS4)*, IEEE;2021, July, pp. 96–100.

48. Speiser JL, Miller ME, Tooze J, Ip E. A comparison of random forest variable selection methods for classification prediction modeling. *Expert Systems with Applications* 2019;134: 93–101.

49. Probst P, Wright MN, Boulesteix AL. Hyperparameters and tuning strategies for random forest. *Wiley Interdisciplinary Reviews: Data Mining and Knowledge Discovery* 2019;9(3): e1301.

50. Sheykhmousa M, Mahdianpari M, Ghanbari H, Mohammadimanesh F, Ghamisi P, Homayouni S. Support vector machine versus random forest for remote sensing image classification: A meta-analysis and systematic review. *IEEE Journal of Selected Topics in Applied Earth Observations and Remote Sensing* 2020;13: 6308–6325.

51. Charbuty B, Abdulazeez A. Classification based on decision tree algorithm for machine learning. *Journal of Applied Science and Technology Trends* 2021;2(01): 20–28.

52. Patel HH, Prajapati P. Study and analysis of decision tree based classification algorithms. *International Journal of Computer Sciences and Engineering* 2018;6(10): 74–78.

53. Rizvi S, Rienties B, Khoja SA. The role of demographics in online learning; A decision tree based approach. *Computers & Education* 2019;137: 32–47.

54. Huang S, Cai N, Pacheco PP, Narrandes S, Wang Y, Xu W. Applications of support vector machine (SVM) learning in cancer genomics. *Cancer Genomics & Proteomics* 2018;15(1): 41–51.

55. Kurani A, Doshi P, Vakharia A, Shah M. A comprehensive comparative study of artificial neural network (ANN) and support vector machines (SVM) on stock forecasting. *Annals of Data Science* 2021;1–26.

56. Battineni G, Chintalapudi N, Amenta F. Machine learning in medicine: Performance calculation of dementia prediction by support vector machines (SVM). *Informatics in Medicine Unlocked* 2019;16: 100200.

57. Abu Alfeilat HA, Hassanat AB, Lasassmeh O, Tarawneh AS, Alhasanat MB, Eyal Salman HS, Prasath VS. Effects of distance measure choice on k-nearest neighbor classifier performance: A review. *Big Data* 2019;7(4): 221–248.

58. Gou J, Ma H, Ou W, Zeng S, Rao Y, Yang H. A generalized mean distance-based k-nearest neighbor classifier. *Expert Systems with Applications* 2019;115: 356–372.

59. Lubis AR, Lubis M. Optimization of distance formula in K-nearest neighbor method. *Bulletin of Electrical Engineering and Informatics* 2020;9(1): 326–338.

60. Chen S, Webb GI, Liu L, Ma X. A novel selective naïve Bayes algorithm. *Knowledge-Based Systems* 2020;192: 105361.

61. Berrar D. *Bayes' theorem and naive Bayes classifier*. Encyclopedia of bioinformatics and computational biology: ABC of bioinformatics. 2018, p. 403.

62. Zhang H, Jiang L, Yu L. Attribute and instance weighted naive Bayes. *Pattern Recognition* 2021;111: 107674.

63. Luque A, Carrasco A, Martín A, de Las Heras A. The impact of class imbalance in classification performance metrics based on the binary confusion matrix. *Pattern Recognition* 2019;91: 216–231.

64. Xu J, Zhang Y, Miao D. Three-way confusion matrix for classification: A measure driven view. *Information Sciences* 2020;507: 772–794.

65. Hasnain, M, Pasha, MF, Ghani, I, Imran, M, Alzahrani, MY, Budiarto, R. Evaluating trust prediction and confusion matrix measures for web services ranking. *IEEE Access* 2020;8: 90847–90861.

# *Index*

Pages in *italics* refer to figures and **bold** refer to tables.